Water in the Universe

Astrophysics and Space Science Library

For other titles published in this series, go to
www.springer.com/series/5664

Arnold Hanslmeier

Water in the Universe

Springer

Prof. Dr. Arnold Hanslmeier
Institut für Geophysik,
Astrophysik und Meteorologie
Universität Graz
Universitätsplatz 5
8010 Graz
Austria
arnold.hanslmeier@uni-graz.at

ISSN 0067-0057
ISBN 978-90-481-9983-9 e-ISBN 978-90-481-9984-6
DOI 10.1007/978-90-481-9984-6
Springer Dordrecht Heidelberg London New York

Library of Congress Control Number: 2010937475

Cover illustration: Water Claimed in Evaporating Planet HD 209458b. Illustration Credit: European Space Agency, Alfred Vidal-Madjar (Institut d'Astrophysique de Paris, CNRS), NASA
Cover design: eStudio Calamar S.L.

Printed on acid-free paper

Springer is part of Springer Science+Business Media (www.springer.com)

Preface

Water is one of the basic elements for life. It is even assumed that the evolution of life is only possible if there is liquid water present. A water molecule has some remarkable properties that make it quite unique in the universe. In the first chapter of this book we will review these basic properties of water and the role of water on Earth. All ancient civilizations realized the importance of water and their cities were constructed near great reservoirs of water. But is water unique on Earth? Do we find water elsewhere in the solar system, on extrasolar planetary systems or in distant galaxies? We will start the search for the presence of extraterrestrial water in our solar system. Surprisingly enough it seems that water in some form and sometimes in only minute quantities is found on any object in the solar system. Even on the planet nearest to the Sun, Mercury, there may be some water in the form of ice near its poles where never the light of Sun heats the surface. And there are objects in the solar system that are made up of a large quantity of water in terms of their mass, like comets and several satellites of the giant planets.

If life depends on the presence of liquid water, there are also places besides Earth where liquid water may be found: beneath the ice crust of several satellites of Jupiter and Saturn there might be hidden a liquid ocean. Such an ice crust provides a shielding against high energetic radiation.

Now, since the first extrasolar planetary systems have been detected, the search for water on such objects has just started. Because from observations it is very difficult to measure the spectroscopic signatures of the atmospheres of such planets, we have to wait for the newly planned observational facilities (both in space and on ground); some of them will be in operation very soon.

Water has been detected almost everywhere on extreme and exotic places in the universe: in 5000 K hot sunspots as well as in cold molecular interstellar clouds. Extreme bright sources can be explained by a MASER mechanism that is based on water molecules. They indicate regions where stars are formed and they can be even detected in galaxies that are at a distance of several 100 million light years.

Water, which consists of hydrogen and oxygen, was formed after the first generation of luminous stars exploded, so it was not present during the first several 100 million years of the history of our universe. This will be reviewed in the last chapter.

The book is intended for the reader interested in astrophysics, astrobiology and science in general. It provides an overview but since more than 350 papers are cited, the reader who wants to go deeper can use these references. It can also be used as a textbook on several topics related to astrobiology.

I want to thank Mr. Ramon Khanna and Mr. Donatas Akmanavičius from Springer for their excellent cooperation. The NASA ADS provides a wonderful tool for searching literature and some introductory remarks are based on information found in the WIKIPEDIA—I want to thank the many unknown authors who contribute to that encyclopedia.

I am also grateful to Dr. Roman Brasja and Prof. Arnold Benz for helpful comments.

Finally, I want to thank my children Roland, Christina and Alina and my girlfriend Anita for their understanding of my scientific passion.

Graz, Austria Arnold Hanslmeier

Contents

List of Tables

Chapter 1
Water on Earth, Properties of Water

In this section we discuss some important chemical properties of water which make this molecule very unique: water has a high surface tension, a high specific heat index and it is the only substance found on Earth at all three states, gas, liquid and solid. There are only few molecules with similar properties like water. The amount of water in the oceans of the Earth is estimated at about 1.4×10^{21} kg. Comparing this with the amount of water estimated on Venus (2×10^{16} kg) we see that Venus is a dry planet, however, there is still a huge quantity of water there.

Moreover, water is present almost ubiquitously throughout the universe. Water molecules have been detected also in the interstellar medium as well in the spectra of stars.

In this chapter we discuss the importance of water on Earth, briefly its role in history and some basic chemical and physical properties of water.

1.1 The Role of Water in History

The Earth is also called the *blue planet* since from space the oceans appear in a dark blue color. In the solar system, the Earth seems to be the only planet where liquid water is present on the surface. Besides as a sources for pure water to drink, big rivers and the sea were used also for transport. Water played also an important factor in the age of industrialization. There is no big city which is not located near a river.

1.1.1 Water in Ancient Cultures

Water is essential for life. This was already recognized by ancient cultures. Humans always settled near sources of water. The ancient Egyptian culture developed near the Nile river and the civilization in Mesopotamia (this is the Greek expression for land between two rivers) flourished between the two rivers Euphrat and Tigris. The Chinese civilization was centered about the Yellow and Yangzi river basins.

A. Hanslmeier, *Water in the Universe*, Astrophysics and Space Science Library 368,
DOI 10.1007/978-90-481-9984-6_1,

Water in Ancient Egypt

Egypt is a very dry region and only the small area around the river Nile can be used for agriculture. The Nile is the heart of the ancient and modern land of Egypt. The floods of the river Nile were more easily predictable than the floods of the Euphrat and Tigris rivers. The cause of the floods are the monsoon-type rains in the Ethiopian highlands. The ancient Egyptians soon recognized that the river started to rise in Egypt at the beginning of July. It reached the flood stage at Assuan by mid of August. Then the flood spread northwards within the next six weeks. The flood covered the floodplain up to a depth of 1.5 m and started to recede by mid of October. Egyptian priests realized that this flood time corresponds to the first time when the star Sirius (they called it Sothis) became visible in the dawn. This heliacal rise of Sirius was an important marker in the Egyptian calendar; the Egyptian year had 365 days. The year was divided into 12 months of 30 days each plus five extra days. The difference between a seasonal year and a civil year of 365 days was 365 days in 1460 years, or about 1 day in 4 years. For more information see also Schaefer, 2000 [293].

Since water is very precious in a region which is dry and mostly a desert, the Egyptians developed a very sophisticated irrigation system. The first water management dates back to 5000 BC in Egypt. The ancient irrigation systems (known as Shadufs, see Fig. 1.1) can still be found along the banks of the Nile river. Today, the water of the Nile is sufficient to more than 70 million people. Since the population of Egypt is growing by about 1 million per year and 96% of that country is desert, Egypt is running out of water.

Water in Ancient Greece

In ancient Greek philosophy, water together with earth wind and fire was one of the four classical elements. To these four elements a fifth element was added known as aether or quintessence (meaning also the void). This fifth element was added by Archimedes (287–212 BC). He believed that stars must be made out of aether because they do not change. Water (symbol ∇) was associated with emotion and intuition.

Thales of Milet (624–546 BC) stated that the origin of all matter in the Cosmos is water. Therefore, he anticipated what was recognized more than 2000 years later by biologists and chemists that water is a prerequisite for life. The four elements air, water, earth and fire were believed to be in a process of continuous change and transformation (Anaximander, 546 BC).

Anaximines (494 BC) claimed that air is the fundamental substance. The soul is air. Fire is rarefied air. When air becomes condensed it becomes first water and then earth. Socrates (469–399 BC) was the mentor of Plato, Aristotle (384–322 BC) was a student of Plato (428–348 BC). Plato associated water with an icosahedron which is formed from twenty equilateral triangles.

There is a nice anecdote about Socrates and a student who asked him how does one gain knowledge. Socrates grabbed the student and pushed his head under the

Fig. 1.1 Egyptian Shaduf near Kom Ombo. Credit: Hajor

water until the student firmly fought to get free again. Socrates told the student, that "When you want a thing as much as you just wanted air, then it will be hard for you not to find it."

Crouch, 1996 [86] discussed the environmental geology of ancient Greek cities, especially water from the Karst springs.

Water in Other Religions and Philosophical Systems

We give few examples of the important role of water as a fundamental element in other religious or philosophical systems: in the Hinduism, the classical elements are bhumi (earth), ap or jala (water), agni or tejas (fire), marut or pavan (air or wind), and byom or akasha (aether). The four elements earth, water, fire and air are also found in Buddhism. In the philosophy of the seven Chakras also water plays an important role: it is called Svadhisthana (Sacral).

In China, the five major planets that were known by ancients are associated with and named after the elements: Venus is gold, Jupiter is wood, Mercury is Water, Mars is Fire, and Saturn is Earth. Additionally, the Moon represents Yin, and the Sun represents Yang.

Water in Ancient Rome

Ancient Rome is famous for its water and wastewater systems and water was considered a deity to be worshipped and most of all utilized in health and art (see also Boni and Boni, 1996 [38]). The availability of huge water supplies was considered a symbol of opulence and therefore an expression of power. The Romans constructed highly sophisticated aqueducts. In their cities, the residents were concentrated around the center. In 312 BC the construction of the first aqueduct to the city of Rome became necessary since the water from local sources and from the river Tiber was not sufficient or became too much polluted by the increasing population. In the aqueducts, the channels were usually rectangular in the cross-section and varied from 0.5 to 2.0 meters in width and from 1.5 to 2.5 meters in depth. Sometimes two or three channels were superimposed, the upper ones being added to the original to accommodate increasing demand. The amount of water that was transported by the Roman aqueducts was up to one million m^3 per day. The per capita use per day of water amounted up to 67 liters. The cloaca maxima was the main wastewater drainage system already built by Etruscan engineers. Interesting aqueduct remains can be admired in Rome, Segovia (Spain), Nimes (France), and Cologne (Germany).

Water Management by Ancient American Cultures

The city of Machu Picchu (see Fig. 1.2) is one of the most important archeological sites in the Americas. It is located in the mountains of Peru and was detected in 1911. It was constructed at a height of 2440 m in 1450 by the Incas for the emperor Pachacuti. The Incas must have calculated the water consumption and concluded that a certain population can be supplied by a spring located there at an elevation of 2458 m. They constructed a water pipeline more than 700 m long. This pipeline supplied up to 300 l/min.

Fig. 1.2 The Inca city of Machu Picchu

Complex systems of irrigation, aqueducts that were constructed by Andean civilizations were investigated by Zimmerer, 1995 [374].

1.1.2 Modern Society and Water

The Greek $y\delta\rho\omega$ means water. Hydrology deals about water resources and the hydrologic cycle on Earth. Also the movement, distribution and quality of water on Earth are part of that subject.

Already in the first century BC M. Vitruvius described the water cycle: precipitation falls in the mountains and this leads to streams and sources.

How is the water distributed on Earth? Most of the water cannot be used for drinking because 97.5% of water on Earth is salt water. 2/3 of fresh water are frozen in polar caps and ice. Most of the rest is underground and only 0.3 percent is surface water. Freshwater lakes, most notably Lake Baikal in Russia and the Great Lakes in North America, contain seven-eighths of this fresh surface water. A typical household uses 65% of fresh water in the bathroom, 15% in the laundry room, 10% in the kitchen and 10% outdoors. This amounts to about 50 l per person per day. The increasing population causes severe problems for water supply. In Fig. 1.3 the estimated scarcity of water for 2025 is shown.

In Table 1.1 a list of countries by freshwater withdrawal mostly based on *The World Factbook* is given.

Today many sophisticated waterway systems are used by the countries. The length of the waterways used by China e.g. exceeds 600 000 km which is about 10 times the length of the waterways in the European Union or the United States.

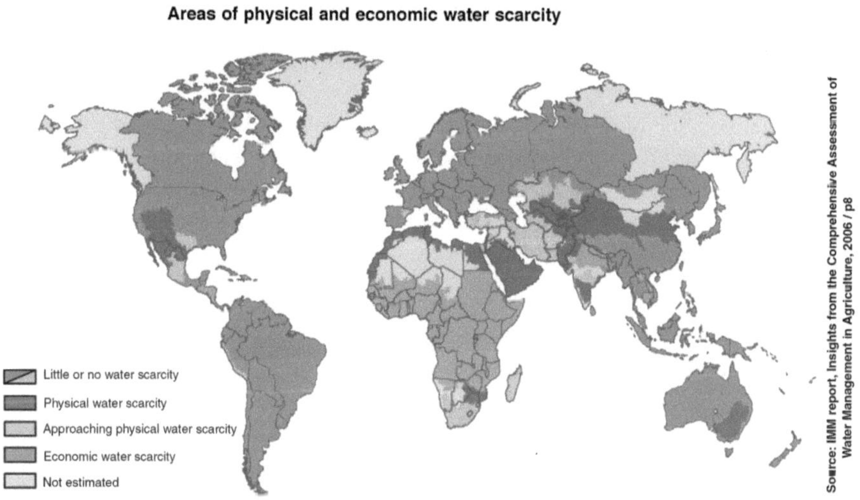

Fig. 1.3 Water scarcity in 2025. Credit: International Water Management Institute

Table 1.1 List of countries by freshwater withdraw. Column T means Total withdrawal (km^3/year), column C Per capita withdrawal (m^3/year), D is the Domestic withdrawal in %, I is the Industrial withdrawal in %, A is Agricultural withdrawal in %

Country	T	C	D	I	A
India	645.84	585	8	5	86
China	549.76	415	7	26	68
Japan	88.43	690	20	18	62
Russia	76.68	535	19	63	18
Egypt	68.3	923	8	6	86
Germany	38.01	460	12	68	20
Spain	37.22	864	13	19	68
France	33.16	548	16	74	10
Australia	24.06	1 193	15	10	75
Saudi Arabia	17.32	705	10	1	89
United Kingdom	11.75	197	22	75	3
Austria	3.67	448	35	64	1

Thompson, 1992 [335], gave a review about water in the Earth's upper mantle. Water is present in all magmas and mantles in the Earth (sometimes only at ppm level). Water lowers the melting temperatures of rocks, decreases the viscosity of magma and thus enhances magma migration. Water may be stored in melts and not in minerals. Such hydrous melts can induce volcanism beneath thick lithosphere.

1.2 The Chemical Elements Water Consists of

A water molecule consists of two chemical elements: Hydrogen, H, which is the most abundant element in the universe, constituting 75% of its mass, and Oxygen, O, which was not produced at the early stages of the evolution of the universe but was formed later in the stellar interiors by fusion reactions. Thus water was not present at the very early phase of the universe.

The composition of water was detected by Lavoisier in 1783. Cavendish demonstrated already in 1766 that hydrogen is one component of water and that hydrogen burns to water. This was a rather revolutionary discovery because since the time of Aristotle, it was believed that water is one of the basic substances. Water is the most abundant molecule on the Earth's surface.

1.2.1 Hydrogen

Hydrogen is the simplest atom, it consists of one positively charged proton and one negatively charged electron in the case of neutral hydrogen, in the case of ionized hydrogen the electron is lost. In the atmospheres of relatively cool stars (also our Sun belongs to that group), there even exists another ion of hydrogen, the H^-, which is a proton surrounded by two electrons. This negative H ion is the main contributor to opacity in cool stars. This H^- gets ionized to H at higher temperatures.

Hydrogen is a colorless, highly inflammable gas. It burns at concentrations as low as 4% in air. When mixed with oxygen it explodes upon ignition. The reaction is the following:

$$2H_2 + O_2 \rightarrow 2H_2O + 286\ \text{kJ/mol} \tag{1.1}$$

Thus the result is water.

Isotopes of Hydrogen

The mass number is the number of protons and neutrons in the nucleus and is written as an upperscript. The atomic number, often denoted by Z, is the number of protons in a nucleus and is denoted by a subscript. Isotopes contain the same number of protons but a different number of electrically neutral neutrons.

There are several isotopes of hydrogen. The most abundant form of hydrogen is ^1_1H. Its nucleus consists of one proton.

The two other isotopes of hydrogen are Deuterium ^2_1H (consists of one proton plus one neutron) and the Tritium ^3_1H (consists of one proton plus two additional neutrons). Thus Deuterium has the atomic number 1 and the mass number 2, Tritium has the atomic number 1 and the mass number 3.

Deuterium is also called heavy hydrogen. It is a stable isotope of hydrogen and its natural abundance is one atom in 6500.[1] Deuterium was detected in 1931 by H. Urey.

Tritium is the radioactive isotope of hydrogen. Its half-life is about 12.32 years and it decays to:

$$^3_1H \rightarrow {}^3_2He + e^- + \bar{\nu}_e \tag{1.2}$$

This is the β^- decay where a neutron decays to a proton and an electron plus an antielectronneutrino:

$$n \rightarrow p + e^- + \bar{\nu}_e \tag{1.3}$$

The decay of Tritium produces ^3He and energy of 18.6 keV is released (5.7 keV is the kinetic energy of the electron, the rest is the energy of the neutrino ν). The decay of tritium produces low energy β radiation.[2] This radiation is not dangerous because it cannot penetrate human skin (only if inhaled or ingested).

Tritium is produced in the Earth's upper atmosphere mainly by the interaction of nitrogen atoms with cosmic ray neutrons n:

$$^{14}_7N + n \rightarrow {}^{12}_6C + {}^3_1H \tag{1.4}$$

[1] This value corresponds to 154 ppm, parts per million.

[2] Because of the production of e^- that are also called β^- particles.

Tritium is also produced e.g. during nuclear weapon explosions. The radioactive properties of tritium are used in research, fusion reactors and also for dim light sources such as e.g. exit signs where it is mixed with phosphor and for the same purpose it is used in watches. Tritium was discovered by E. Rutherford, M.L. Oliphant, and P. Harteck, in 1934.

Hydrogen in the Universe

As it was already mentioned, hydrogen was formed during the earliest stages of the universe when the temperature became low enough due to adiabatic expansion so that protons could form. Neutral hydrogen formed when electrons were captured by the protons. About 100 s after the Big Bang, Deuterium formed and for another few minutes fusion of hydrogen into helium created the primordial chemical composition of the universe.

Hydrogen is the most abundant element in the universe, 75% of normal matter is made up of this gas by its mass, 90% by the number of atoms. Hydrogen is found nearly everywhere, in stars, planets and interstellar matter. Hydrogen molecules, H_2, form molecular clouds that play a key role in the star formation and are therefore studied intensively by astrophysicists. Giant molecular clouds contain 10^4 to 10^6 times the mass of the Sun, $M_\odot = 2 \times 10^{30}$ kg, a diameter of about 100 ly[3] and, though they often have a very spectacular appearance, the density is very low, 10^2 to 10^3 particles per cm^3. Such molecular clouds are regions of star formation. An example of a nearby star forming region is the Orion nebula (Fig. 1.4) at a distance of about 1200 ly.

1.2.2 *Oxygen*

Oxygen (from the Greek $o\xi y\sigma$, oxys, sharp, acid) was independently discovered by J. Priestley in Wiltshire, in 1774, and C.W. Scheele, in Uppsala, a year earlier, but Priestley is usually given priority because he published his findings first.

Oxygen, O, is a highly reactive, non-metallic element. Its compounds that are formed with nearly all other chemical elements are called oxides. At standard temperature and pressure[4] two atoms of oxygen form a molecule of O_2 which is a diatomic gas. Oxygen is a colorless odorless gas at standard temperature. The atomic number is 8.

Ozone O_3 is produced in the upper Earth atmosphere. Solar UV radiation splits O_2 molecules. Then a recombination of O with other O_2 molecules occurs.

---

[3]1 Light year ly denotes the distance that light travels within one year, 1 ly $= 10^{13}$ km.

[4]Standard temperature and pressure (denoted as STP) refers to nominal conditions in the atmosphere at sea level. Standard temperature is 0°C, standard pressure denotes 760 mm in mercury barometer.

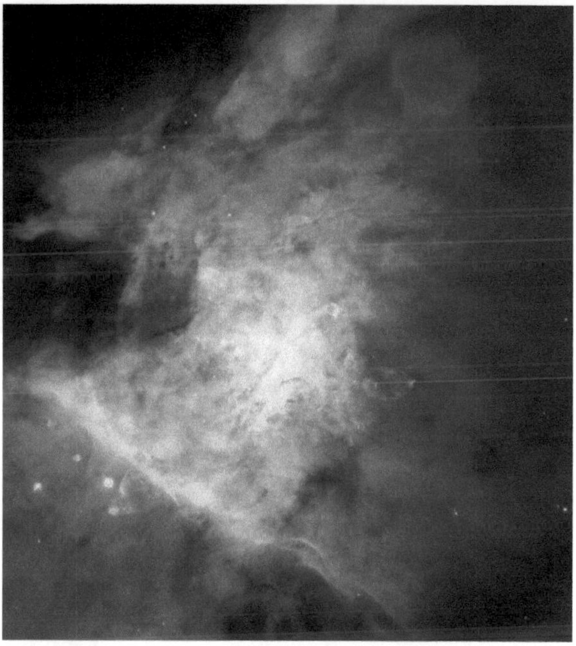

Fig. 1.4 The Orion nebula consists mainly of hydrogen and is a typical star forming region, the youngest stars are only 300 000 years old which is less than 1/100 000 the age of the solar system. Credit: Hubble Space Telescope, HST

These are the well known Chapman reactions:

$$O_2 + h\nu \rightarrow O + O \quad \lambda = 180\text{–}240\ \text{nm} \tag{1.5}$$

$$O + O_2 + M \rightarrow O_3 + M \tag{1.6}$$

$$O_3 + h\nu \rightarrow O + O_2 \quad \lambda = 250\ \text{nm} \tag{1.7}$$

$$O + O_3 \rightarrow O_2 + O_2 \tag{1.8}$$

M is an inert species which absorbs excess energy from the excited O molecule and h denotes the Planck constant (6.626×10^{-34} J s).

The isotopes of oxygen are $^{16}_{8}O$ (8 neutrons), $^{17}_{8}O$ (9 neutrons), $^{18}_{8}O$ (10 neutrons), the most abundant is $^{16}_{8}O$ with an abundance of 99.762%. The other isotopes of oxygen are unstable: $^{15}_{8}O$ with a half-life of 122.4 s, $^{14}_{8}O$ with a half-life of 70 s and even less stable isotopes.

Oxygen is soluble in water. However, the solubility is temperature dependent. In Fig. 1.5 the solubility of oxygen in fresh water is given, for sea water the curve is parallel but shifted to lower values.

Oxygen and Palaeoclimatology

Water consists of two hydrogen and one oxygen atom. The oxygen atom could be a lighter or a heavier isotope (see Fig. 1.6). The evaporation of water depends on temperature.

- Water molecules containing the lighter isotope are slightly more likely to evaporate.

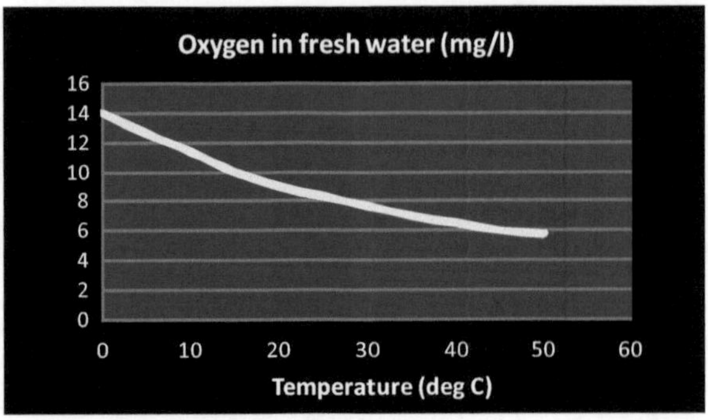

Fig. 1.5 Solubility of oxygen in fresh water

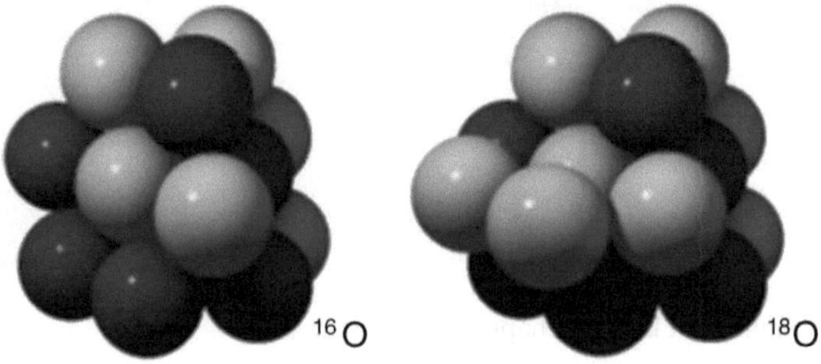

Fig. 1.6 Two isotopes of oxygen: ^{16}O and ^{18}O. From the ratio of these isotopes, the climate in the past can be reconstructed

- Water molecules containing the heavy isotope condense more readily and fall as precipitation.

Therefore, fresh water and polar ice contains less of the heavy ^{18}O isotope (0.198%) than air (0.204%) or seawater (0.1995%). By comparing the isotope ratio of water at a depth of 200–500 m with the isotope ratio measured in ice cores, the palaeoclimate can be reconstructed.

The long-term record of oxygen isotopes in the Greenland Ice Sheet extends back more than 250 000 years.

Since the ^{18}O includes contributions not only from sea surface temperature but also global ice volume and local salinity, Rostek et al., 1993 [282] reconstructed sea surface temperature and salinity in the juncture of the Arabian Sea and the Bay of Bengal combining ^{18}O and alkenone records. Alkenones are highly resistant organic compounds produced by phytoplankton. They were able to reconstruct the Indian monsoon over the last 170 kyr finding large variations in the monsoon during the transition from the last glacial period (e.g. in European Alps the Würm glacial period lasted from 12–110 kyr BP) to the present interglacial.

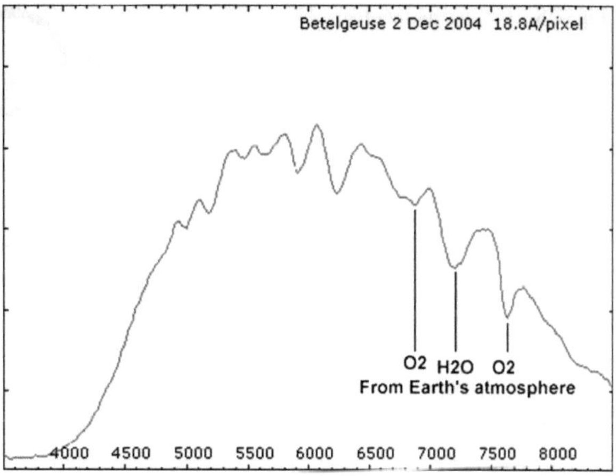

Fig. 1.7 Terrestrial water molecule line and oxygen lines in the spectrum of the supergiant star Betelgeuse. On the x-axis the wavelength is given in Å (1 Å = 0.1 nm). Credit: Robin Leadbeater

Oxygen in the Universe

Oxygen is synthesized at the end of stellar evolution by fusion processes. It is the third most abundant chemical element in the universe, e.g. 0.9% of the Sun's mass is oxygen. The Earth's atmosphere contains about 10^{15} tons of oxygen (21.0% by its volume, 23.1% by its mass). The oxygen content of Mars's atmosphere amounts only 0.1%.

On Earth, photosynthesis releases oxygen in the atmosphere while respiration and decay remove it. At present there is an equilibrium, 1/2000 of the entire oxygen content in the atmosphere is produced and consumed per year.

Oxygen absorption lines can be observed e.g. at 687 and 760 nm (see also Fig. 1.7). It has been proposed to use such measurements as biomarkers for life on extrasolar planets.

1.3 Water, Chemical and Physical Properties

1.3.1 Chemical Properties

Water has some almost unique chemical properties which can be explained by the structure of its molecules.

Water molecules are an example of a polar covalent bond. Electrons are unequally shared between the hydrogen and oxygen atom. Oxygen has six valence electrons and would need eight to complete its valence shell. Therefore, it shares two electrons from two hydrogen atoms which become positively charged. The large oxygen atom has a stronger affinity for electrons than the small hydrogen atoms. The bond angle is 104.5° (tetrahedral). The molecule is bent, both hydrogen atoms are on the same side (Fig. 1.8). The bent shape of a water molecule can be explained as follows:

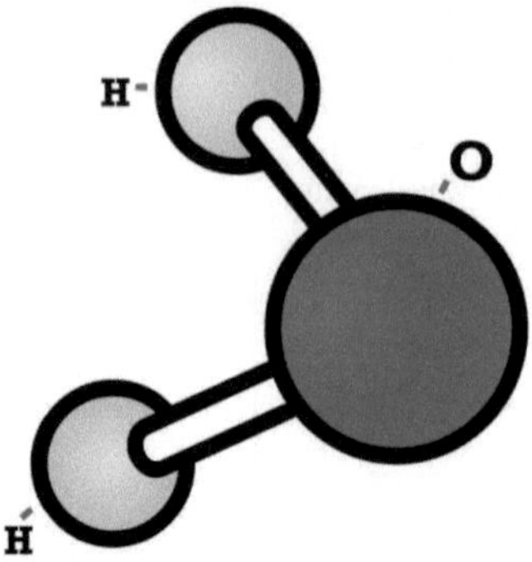

Fig. 1.8 A water molecule

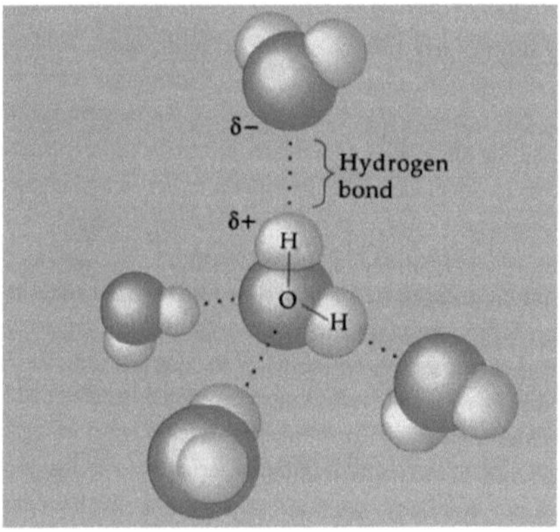

Fig. 1.9 Hydrogen bonds in water

- Tetrahedral arrangement around the oxygen
- Ion pairs.

The charge separation induces temporary dipoles even in non-polar molecules. This explains why water is an excellent solvent. Water can solve and bring together a variety of substances that are the basis for life.

If the bond were linear, this would have a profound effect for life on Earth. Pure water has a neutral pH of 7; it is neither acidic nor basic.

Considering a water molecule, the side with the hydrogen atom (positive charge) attracts the oxygen side (negative charge) of a different water molecule. This attraction between two water molecules is also called hydrogen bond (Fig. 1.9).

1.3.2 Physical Properties of Water

At temperatures found normally on Earth, water exists in all three states: liquid, solid and gas. It has some unique properties.

- Under normal pressure conditions, water freezes at 0°C and boils at 100°C which defines the Celsius temperature scale.

Table 1.2 Boiling point of water as a function of pressure

Pressure (Pa)	Boiling point (°C)	Pressure (bar)	Boiling point (°C)
3000	26.4	190 000	119
7000	38.7	250 000	127
14 000	52.2	300 000	134
21 000	60.8	330 000	137
28 000	67.2	400 000	144
34 000	72.3	500 000	152
41 000	76.7	610 000	159
66 000	83.8	690 000	164
76 000	92.1	1 000 000	181
90 000	96.6	2 100 000	214
100 000	100	2 900 000	233
110 000	101	4 000 000	250
120 000	104	5 000 000	264
150 000	112	6 000 000	276

- In its solid form, ice, it is less dense than in its liquid form. This is again essential for life because ice always forms at the surface of a lake, and it swims.
- Specific heat: This is a measure of the heat energy required to increase the temperature of a substance by a certain temperature interval. The heat energy is measured in Joule and is defined by:

$$\Delta Q = mc\Delta T \tag{1.9}$$

Here, m is the mass of the substance, c the specific heat and ΔT the temperature difference. The boiling point of a liquid is the temperature at which the vapor pressure equals the environmental pressure surrounding the liquid. The SI unit of pressure is N/m^2 which is called Pascal, Pa. The standard pressure is the pressure at 1 bar (100 kPa, this is the current IUPAC definition).[5] The standard reference conditions are: temperature 0°C, pressure 100 kPa.

The boiling point increases with increased pressure up to the critical point, where the gas and liquid properties become identical. At standard pressure, water boils at 100°C, on the top of Mount Everest, the pressure is 260 Pa and the boiling point of water is 69°C.

In Table 1.2 the boiling point of water is given for different values of the surrounding pressure. From that table it is seen that at a very low pressure of 0.07 bar, water already boils at 38.7°C. This is about ten times the present pressure in the Martian atmosphere. Therefore, no liquid water can exist on the surface of Mars at present, because of the low pressure there. The boiling point of water at low pressure is listed in Table 1.3.

[5] International Union of Pure and Applied Chemistry.

Table 1.3 Boiling point of water at low pressures

Temperature (°C)	Pressure (Pa)
0	611
5	872
10	1228
12	1403
14	1599
16	1817
17	1937
18	2064
19	2197
20	2338
21	2486
22	2644
23	2809
24	2984
25	3168

Table 1.4 Specific heat capacity of water and some other elements and substances

Compound	Heat of vaporization ($kJ\,mol^{-1}$)	Heat of vaporization ($kJ\,kg^{-1}$)
Ammonia	23.35	1371
Butane	21.0	362
Ethanol	38.6	841
Hydrogen	0.46	451.9
Methane	8.19	510
Methanol	34.5	1104
Propane	15.7	356
Phosphine	14.6	429.4
Water	40.65	2257

In Table 1.5 the specific heat capacity of water and some other elements is given.

In Table 1.4 the heat of vaporization is given for several substances—some of them play an important role in the chemistry of giant planets. The heat of vaporization is the amount of heating for turning a certain amount of a liquid into a vapor at its boiling point, without a rise in temperature of the liquid.

From this table we see that water has a high heat capacity.

The maximum density is found at a temperature of 4°C. The latent heat is the amount of energy in the form of heat released or absorbed by a substance during a change of phase (e.g. from solid to liquid). Values for the specific latent heat for different substances and phase transitions are given in Table 1.6.

Table 1.5 Specific heat capacity of water and some other elements and substances

Element, substance	$J\,g^{-1}\,K^{-1}$
Hydrogen	14.30
Gasoline	2.22
Ammonia	4.70
Oxygen	0.92
Glass	0.84
Carbon dioxide	0.84
Copper	0.39
Concrete	0.88
Water (liquid at 25°C)	4.1813 = 1 cal
Water (steam)	2.08
Water (ice)	2.05

Table 1.6 Latent heats and change of phase temperatures for some substances

Substance	Latent heat fusion J/g	Melting point °C	Latent heat vaporization J/g	Boiling point °C
Water	334	0	2275 (at 0°C)	100
Alcohol, ethyl	108	−114	855	78.3
Ammonia	339	−75	1369	−33.34
Carbon dioxide	184	−57	574	−78
Hydrogen	58	−259	455	−253

The heat of vaporization (evaporation) is the energy required to transform a substance into a gas and is also known as the enthalpy of vaporization ($\Delta_v H$). The enthalpy of condensation is exactly equal to the enthalpy of vaporization, but has the opposite sign. Why do have substances different enthalpy? From the molecular point of view the enthalpy is the energy that is needed to overcome the Van der Waals force between the atoms of a substance. These forces are very weak e.g. for Helium, however for water, because of the hydrogen bonds explained in the previous section, the forces are much stronger. The enthalpy of condensation is the heat that is released to the surroundings to compensate for the drop in entropy. As a gas condenses to a liquid the entropy drops; if T_b denotes the boiling point, then this change of entropy may be written as:

$$\Delta_v S = S_{\text{gas}} - S_{\text{liquid}} = \Delta_v H / T_b \quad (1.10)$$

For more details on these thermodynamical subjects the reader should consult textbooks.

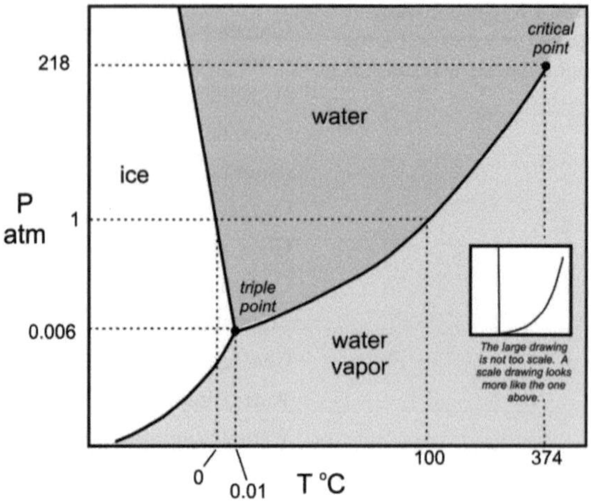

Fig. 1.10 Water: phase diagram. The triple point denotes the pressure and temperature where water can coexist in all three states. Credit: SERC at Carleton College

1.3.3 Evaporation and Condensation

Consider a quantity of water. Whenever a water molecule has enough energy to leave the surface of water, we speak of evaporation. It is also evident, that by this process, since the evaporating molecule takes the energy with it, the remaining water becomes cooled—there is less energy left there. This effect is known as evaporative cooling and a well known process is perspiration.

Such a process can be also found on the dwarf planet Pluto. Infrared measurements made with the Submillimeter Array in Hawaii have shown that Pluto has about 10 degrees lower surface temperature than its satellite Charon. Sunlight causes the nitrogen ice on the surface of Pluto to sublimate (phase transition from frozen to gas) which causes a cooling.

The dew point is the temperature to which air must be cooled for water vapor to condense into water. The condensed water is then called dew, the dew point is a saturation point. When the dew point falls below freezing, the water vapor creates frost (frost point). Humidity is the amount of water in air. Relative humidity is the ratio of the partial pressure of water vapor in a parcel of air to the saturated vapor pressure of water vapor at a given temperature. Absolute humidity is the quantity of water in a particular volume of air and it changes as air pressure changes.

When water vapor condenses onto a surface, this surface will be warmed and the surrounding air cooled. In the atmosphere, condensation produces clouds, fog and precipitation. Water vapor will condensate also on surfaces, when the temperature on that surface is below the dew point temperature of the atmosphere. Water vapor is lighter or less dense than dry air and therefore it is buoyant with respect to dry air. The molecular mass of water is 18.02 g/mol, the average molecular mass of air (79% N_2, 21% O_2) is 28.57 g/mol. Avogadro's law states that at standard temperature and pressure the molar volume of a gas is 22.414 l/mol. From that we can calculate the density $\rho = m/V$ and find the values: water vapor 0.8 g/l, dry air 1.2 g/l.

Figure 1.10 is the phase diagram of water. This shows at which temperatures and pressures water is found in which of the three states: gas, liquid or solid.

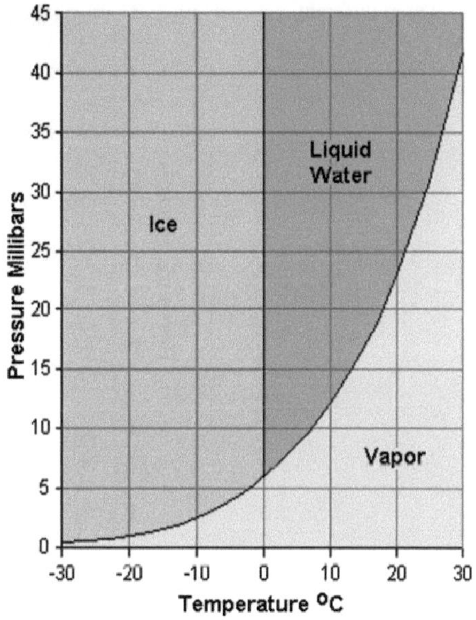

Fig. 1.11 Water: phase diagram

Figure 1.11 shows the exact values of pressure and temperature where water can be found in either liquid, solid or gaseous state. At a pressure of 5 mbar and a corresponding temperature of 0°C water can coexist in all three states. On Mars, because of the low atmospheric pressure below 10 mbar and temperatures below 0°C, there is no liquid state of water possible. If the pressure in the Martian atmosphere doubles to e.g. 20 mbar then liquid water could exist between the temperature interval from 0°C to 20°C.

1.3.4 Ice

When liquid water is cooled below 0°C (273.15 K) a phase transition to ice occurs at standard atmospheric pressure. This type of water ice is also called I_h. Frost is a deposit from a vapour with no intervening liquid phase. Light reflecting from ice can appear blue, because ice absorbs more of the red frequencies than the blue ones.

Ice appears in nature in forms as varied as snowflakes and hail, icicles, glaciers, pack ice, and entire polar ice caps. An unusual property of ice frozen at a pressure of one atmosphere is that the solid is some 9% less dense than liquid water.

- At 0°C ice has a density of 0.9167 g/cm^3, water has a density of 0.9998 g/cm^3.
- At 4°C liquid water has a density of 1.000 g/cm^3.

Water becomes less dense at the transition to ice because the water molecules begin to form hexagonal crystals of ice. Ice is the only known non-metallic substance to expand when it freezes. The density of ice increases slightly when the temperature decreases. For example at a temperature of -180°C the density of ice is 0.934 g/cm^3. Ice can also be superheated beyond its melting point.

Fig. 1.12 Water: different states of ice

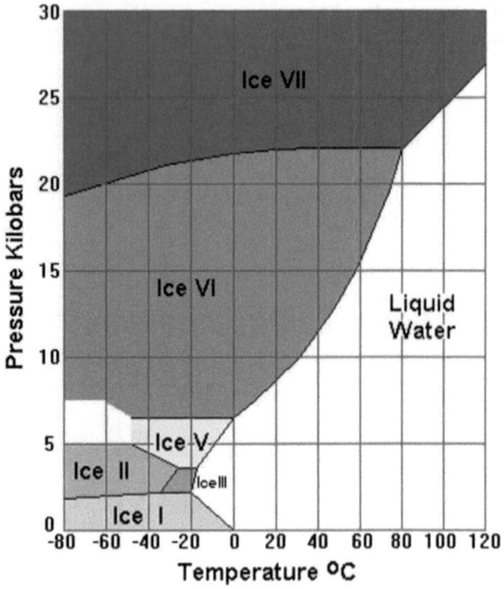

There exist different types of ice (see also Fig. 1.12). The ice we know from everyday live (also snow) has a hexagonal structure. At higher temperatures and pressures ice can also form a cubic structure I_c). Other forms of ice are called II, III, V, VI, VII, VIII, IX and X. The difference between these forms is their crystalline structure. One also speaks of low-density amorphous ice (LDA), high-density amorphous ice (HDA), very high-density amorphous ice (VHDA) and hyperquenched glassy water (HGW).

The predominant form of ice found on Earth is hexagonal crystalline ice. The ice found on extraterrestrial objects (e.g. comets) is amorphous. If hexagonal ice is found on some other planet or satellite of a planet its formation is explained by volcanic action.

In history ice played also an important role for cooling. Let us give few examples: Until recently, the Hungarian Parliament building used ice harvested in the winter from Lake Balaton for air conditioning. Icehouses were used to store ice from winter. In 400 BC in Persia ice was brought in during the winters from nearby mountains in bulk amounts, and stored in specially designed, naturally cooled refrigerators, called yakchal.

Amorphous Ice

Amorphous ice is lacking a crystal structure. There are three different forms of amorphous ice:

- low-density (LDA) formed at atmospheric pressure, or below,
- high density (HDA) and
- very high density amorphous ice (VHDA), forming at higher pressures.

LDA forms by extremely quick cooling of liquid water ("hyperquenched glassy water", HGW), by depositing water vapour on very cold substrates ("amorphous solid water", ASW) or by heating high density forms of ice at ambient pressure ("LDA").

Ice I_h

Normal hexagonal crystalline ice. Virtually all ice in the biosphere is of this type.

Ice I_c

A metastable cubic crystalline variant of ice. The oxygen atoms are arranged in a diamond structure. It is produced at temperatures between 130–150 K, and is stable for up to 200 K, when it transforms into ice I_h. Occasionally this type of ice is found in the upper atmosphere.

For properties of the other forms of ice see the relevant textbooks.

1.3.5 H_2O^+

This ionized form of a water molecule is sometimes also called the fourth state of water. The formation process is quite complex, basically there exist two mechanisms. Generally, H_2O^+ originates from water molecules colliding with H^+, the ionized hydrogen atom, or from OH^+ colliding with a hydrogen molecule.

- Interstellar dust, catalytic reaction. One oxygen atom and two hydrogen atoms combine, they are frozen and then they start to evaporate.
- If gas is radiated by far UV or X-rays CO splits up and at the temperature $T > 250$ K the oxygen reacts with H_2 and forms OH^+. This molecule reacts with H_2 and H_2O^+ is formed.

Both reactions occur in star forming regions. These reaction take place in star and planet forming gas irradiated by far UV or X-rays. The forth form of water has been detected outside the solar system by the satellite mission Herschel Space Observatory with the HIFI instrument (a high resolution spectrometer for the far IR).[6] For example Bonev et al., 2002 [37] measured H_2O^+ in the plasma tails of comets at a wavelength of 615.886 nm. Further earlier observations of H_2O^+ in other comets (e.g. Halley's comet) are given in that paper.

[6] Benz, A., 2010, private communication.

1.4 Chemical Reactions and Water

1.4.1 Chemical Bonds

There exist two types of chemical bonds. When ions with opposite charges form a compound there occurs electrical attraction holding the ions together and this is called ionic bond. For example a hydrogen atom can give up its sole electron and a hydrogen ion, H^+ is formed, Chlorine for example gains electrons forming chlorine ions, Cl^-.

When atoms form bonds by sharing electrons we speak of a covalent bond. An example of a covalent bond is molecular hydrogen, H_2. Carbon can form covalent bonds simultaneously with four other atoms and therefore complex molecules such as sugars, proteins and others are created.

When an atom gives up electrons, we say it is oxidized, when an atom gains electrons it is reduced. We gain energy from food by oxidation of sugar and starch molecules.

Forming bonds requires energy, breaking bonds generally releases energy (some activation energy is needed).

1.4.2 Acids and pH Value

Acids and bases are substances that behave differently in water: Acids give up hydrogen ions, H^+. Hydrochloric acid dissociates in water to form H^+ and Cl^- ions. Substances that release hydroxide ions, OH^- are called bases. For example sodium hydroxide, NaOH dissociates into OH^- and Na^+. Acids in the stomach dissolve food, acids in soil help make nutrients available to growing plants.

The strength of an acid and base is given by the pH value. This value denotes the negative logarithm of the H^+ concentration. Pure water has a pH of 7. The H_3O^+ is called a hydroxonium ion and is responsible for the acidic properties of the solution. For example:

$$CH_3COOH + H_2O \Longleftrightarrow CH_3COO^- + H_3O^+ \tag{1.11}$$

On the other hand, the hydroxide ion OH^- is responsible for the alkaline properties of the solution.

The sulfuric acid is the major component of the clouds of the atmosphere of Venus.

1.4.3 Hydrates, Water in Crystals

In chemistry the term hydrates denote substances that contain water. The water molecules are either bound to a metal center or crystallized with the metal complex.

Water of crystallization is water that occurs in crystals but is not covalently bonded to a host molecule or ion. An example of such a structure is $NiCl_2(H_2O)_6$. Thus $1/3$ of the water molecules in the crystal are not directly bonded. Proteins crystallize with unusual large amounts of water (more than 50%) in the crystal lattice.

Salts are compounds composed of a metal ion plus a non metal (or polyatomic) ion, e.g., sodium chloride (NaCl), and sodium phosphate (Na_3PO_4).

Hydrated salts (or Hydrates) are salts which have a definite amount of water chemically combined. Some common hydrates are:

$$CuSO_4 \cdot 5H_2O$$

This is called a copper (II) sulphate pentahydrate. The dot indicates an attractive force between the polar water molecules and the positively charged metal ion. On heating, the attractive forces are overcome and the water molecules are released

$$CuSO_4 \cdot 5H_2O \rightarrow CuSO_4 + 5H_2O$$

In this example the blue hydrated copper (II) sulphate is transformed after heating into the white anhydrous copper (II) sulphate.

The water released on heating is called the water of hydration.

For example stabilities of pure rock-forming hydrous silicates on Venus' surface as a function of elevation were investigated by Zolotov, Fegley and Lodders, 1997 [376].

1.4.4 Water: Spectral Signatures

Water absorbs longer wavelength stronger than shorter wavelengths. The reflectance of shorter wavelengths is higher and this is the reason why water looks blue or blue-green in the visible and darker when observed at IR wavelengths.

Since water in nature is not pure, there are always some suspended particles in the upper layers of a water body. The reflectivity of water increases and it appears brighter the color being shifted slightly towards longer wavelengths.

Chlorophyll in algae absorbs more of the blue wavelengths and reflects the green, making the water appear more green in color when algae is present. The topography of the water surface (rough, smooth, floating materials, etc.) can also lead to complications for water-related interpretation due to potential problems of specular reflection and other influences on color and brightness (see also Fig. 1.13). The reflectance of clear water is generally low. However, the reflectance is maximum at the blue end of the spectrum and decreases as wavelength increases. Hence, clear water appears dark-bluish. Turbid water has some sediment suspension which increases the reflectance in the red end of the spectrum, accounting for its brownish appearance. The appearance of terrestrial water lines in the spectra of stars is demonstrated in Fig. 1.7.

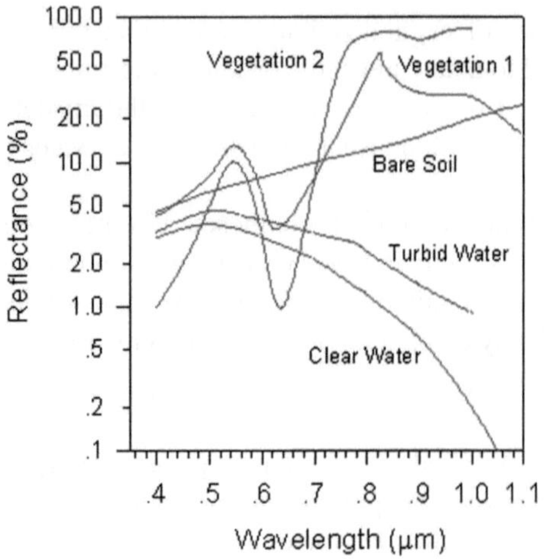

Fig. 1.13 Spectral irradiance of water. Credit: http://www.crisp.nus.edu.sg/

1.5 The Hydrologic Cycle

1.5.1 Evaporation and Precipitation Balance

The hydrologic cycle is very familiar to us and describes the balance between evaporation and precipitation thus the path of water through our environment. The oceans cover about 70 percent of the earth's surface.

Let us consider this balance in more detail:

- *Evaporation*: The evaporation from the oceans is about 400 000 km^3/yr. The evaporation from soil, streams, rivers and lakes is less than a tenth of this value: 30 000 km^3/yr. Transpiration from vegetation is higher than the latter value: 41 000 km^3/yr.
- *Precipitation*: The precipitation over ocean is 385 000 km^3/yr, precipitation over land is 111 000 km^3/yr. Again precipitation over ocean is much higher than over land.
- An important factor for weather and climate is the transport of moist air from ocean to land: 40 000 km^3/yr. Thus about one-tenth of water evaporated from oceans falls over land. This water is recycled through terrestrial systems and drains back to the oceans in rivers.
- Finally percolation through porous rock and soil to groundwater has to be taken into account. About 40 000 km^3 are carried back to the oceans each year.

Therefore 90 percent of the water evaporated from the ocean falls back on the ocean as rain. But considering the numbers given above we see that there is a surplus of water on land. There exist two additional "sinks" of water: part of it is incorporated into biological tissues of living organisms and the greater part of it seeps into the ground where it may be stored for a while (from days to several thousands of years) as soil moisture or groundwater.

Table 1.7 Earth's water compartments. The estimated volume is given in 10^3 km^3, the percentage in % total water and finally the average residence time, that is the time when a molecule from the compartments undergoes a hydrologic cycle again

Compartment	Volume	%	Res. time
Total	1.4×10^6	100	2800 years
Ocean	1.37×10^7	97.6	3000–30 000 years
Ice and snow	29 000	2.07	1–16 000 years
Groundwater down to 1 km	4000	0.28	days—thousands of years
Lakes and reservoirs	125	0.009	1 to 100 years
Saline lakes	104	0.007	10 to 1000 years
Soil moisture	65	0.005	2 weeks to a year
Biol. moisture in organisms	65	0.005	1 week
Atmosphere	14	0.001	8 to 10 days
Swamps and marshes	3.6	0.003	months to years
Rivers and streams	1.7	0.0001	10 to 30 days

The presence of water on our planet regulates the climate and is essential for life. Oceans store heat and release it slowly. Wind currents distribute heat in the latent energy of water vapor. In the tropics warm, humid air rises and is transported to cooler latitudes.

In Table 1.7 an overview of the earth's water compartments is given. It is seen that only 2.4 percent of water on Earth is found outside of the oceans which contain more than 97 percent of all the liquid water. Note that the water of crystallization in rocks is far larger than the amount of liquid water! Oceans contain also 90 percent of the living biomass on Earth. On the average, an individual molecule spends about 3000 years in the ocean before it evaporates and starts through the hydrologic cycle again. In deep ocean trenches there is almost no exchange between water molecules and therefore they may remain undisturbed for tens of thousands of years.

There are estimates that the water in the deep lithosphere is of the amount of 27×10^{18} tons or 94.7% of the global total.

As it has been stressed already, oceans strongly influence the climate by storing heat. Moreover, there are also currents that transport warm water from the equator to higher latitudes and cold water from the poles to the equator. The Gulf Stream flows from the coast of North America toward northern Europe. at a flow rate of 10–12 km/hour. This current carries 100 times more water than all rivers on earth put together. In tropical seas the water is warmed by the sun, diluted by rainwater and aerated by waves. In higher latitudes surface waters are cold and more dense. Those dense waters sink to the bottom to the ocean floors flowing toward the equator, warm water however is less dense and floats on top of this cold water. Also different salinity plays an important role in these processes.

Glaciers, ice caps and snowfields tie up 90 percent of the fresh water (which makes 2.4 percent of liquid water). During the last ice age, 18 000 years ago, about one-third of the continental landmass was covered by ice sheets and since then most of this ice has melted. The largest remnant is in Antarctica, here the ice sheets can be as much as 2 km thick and 85 percent of all ice on earth is stored there. The smaller ice sheet on Greenland and the floating ice around the North Pole makes 10 percent of the ice and the mountain snow peaks and glaciers constitute the remaining 5 percent.

Glaciers are in fact rivers of ice sliding very slowly downhill. Polar ice sheets and alpine glaciers contain more than three times as much fresh water as all the lakes, ponds, streams, and rivers.

1.5.2 The Hydrologic Cycle and Climate Change

Tiny particles called aerosols, are released by human activities into the atmosphere. These anthropogenic aerosols enhance scattering and absorption of solar radiation. As a consequence

- less radiation reaches the Earth's surface,
- a heating of the atmosphere,
- brighter clouds are produced that release less precipitation,
- changes in the atmospheric temperature,
- suppression of rainfall,
- less effective removal of pollutants,

occur and the hydrological cycle becomes weaker with severe implications to the availability and quality of fresh water (Ramanathan et al., 2001 [271]). Constraints on future changes in climate and the hydrologic cycle were studied by Allen and Ingram, 2002 [5]. Climate change is expected to accelerate water cycles and thereby increase the available renewable freshwater resources, therefore this would slow down the increase of people living under water stress (currently more than 2 billion people, see Oki and Kanae, 2006 [252]). In Yang et al., 2003 [369] it is demonstrated that while both the radiative heating by increasing CO_2 and the resulting higher sea surface temperatures contribute to warm the atmosphere, they act against each other in changing the hydrological cycle. As a consequence, in a warmer climate forced by increasing CO_2 the intensity of the hydrological cycle can be either more or less intense depending upon the degree of surface warming. For a recent review on that topic see also Wild and Liepert, 2010 [357] where further references can be found.

Chapter 2
Life and Water

Water is an essential element for life. Life, especially extraterrestrial life is discussed in many textbooks. Astrophysical and astrochemical insights into the origin of life were reviewed by Ehrenfreund et al., 2002 [114] and Chyba and Hand, 2005 [68]. In this chapter we will outline how life can be defined and has evolved on Earth and the role of water for this process.

2.1 Life and Environment

2.1.1 The Importance of Water

In living organisms, water has a number of roles:

- solvent
- temperature buffer
- metabolite
- living environment
- lubricant—to minimize friction (e.g. synovial fluid is encapsulated within the joints).

The water amount in the human body constitutes about 60% of the body weight. Of this, 20% is extracellular body fluid which is made of 5% plasma and 15% tissue fluid. The tissue fluid and the plasma are in a steady state with the fluid inside the cells. There is a strict balance between water intake and water losses (homeostasis). In Fig. 2.1 the balance between daily water intake and water losses for the human body is given in percentages (from a total of 2.5 litres) per 24 hours.

2.1.2 Definition of Life

What are the differences between life and matter? This question appears very simple, however it is not. A philosophical review on that question is given by Gayon,

A. Hanslmeier, *Water in the Universe,* Astrophysics and Space Science Library 368,
DOI 10.1007/978-90-481-9984-6_2, © Springer Science+Business Media B.V. 2011

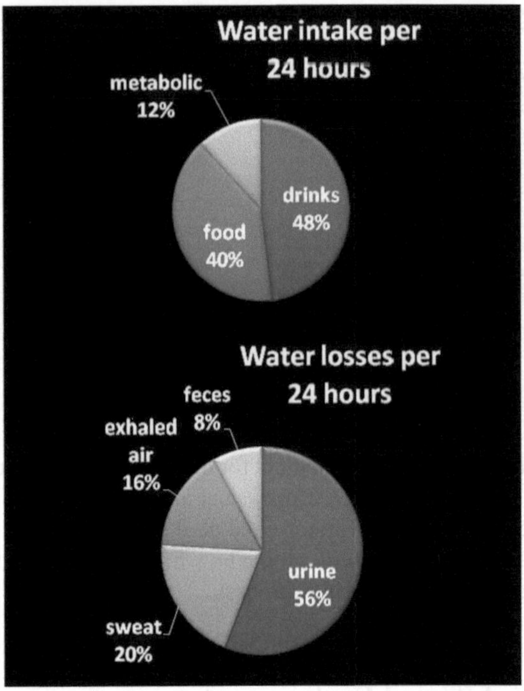

Fig. 2.1 The balance between water intake and water losses in the human body

2010 [138]. Aristotle defined life as animation, life can be also defined as mechanism or organization (Kant). Until very recently, viruses were not considered in discussions on the origin and definition of life. This situation is rapidly changing, and it has been recognized that viruses have played (and still play) a major innovative role in the evolution of cellular organisms (Forterre, 2010 [134]). Life scientists and chemists have not come to a conclusive definition of life. There are several characteristics for life as we know it from Earth:

- *Cells*: all living organisms consist of cells, there exist unicellular and multicellular organisms. A typical cell size is 10 μm; a typical cell mass is 1 nanogram. Cells mainly consist of cytoplasm bound by a very thin membrane. This membrane serves also as a protection against the environment. There exist basically two types of cells: prokaryotic and eukaryotic. The prokaryotic cells are simpler than the eukaryotic, they contain no nucleus. Bacteria are prokaryotic. The first cells appeared on Earth up to 3.8 billion years ago (see e.g. Mojzsis et al. 1996 [234]).
- *Growth and reproduction*: when organisms reproduce, the offspring resemble the parents. Cells must be able to divide. There are two mechanisms, the *mitosis* and the meiosis. For example, the human body loses about 50 million cells each seconds that must be replaced. In the mitosis genetic information is equally distributed to two daughter nuclei. During the crossing over, the chromosomes are aligned in the cell's equatorial plane, the DNA gets replicated. Each new cell gets one of the two daughter nuclei. In the process of *meiosis* which is essential for sexual reproduction and might have appeared first 1.4 billion years ago, one par-

ent cell produces four daughter cells (autocatalytic replication). Daughter cells contain half the number of chromosomes found in the original parent cell and with crossing over, are genetically different.

- *Response to stimuli*: a major characteristic of all living things. Plant responses to stimuli are generally much slower than those of animals.
- *Metabolism*: collective product of all the biochemical reactions taking place within an organism. New cytoplasm is produced, damage repaired and normal cells are maintained. Metabolism includes photosynthesis, respiration, digestion and assimilation. All these include complex chemical processes.
- *Movement*: also plants can move. Leaves of sensitive plants (e.g. Mimosa pudica) fold within a few seconds after being disturbed or subjected to sudden environmental changes.
- *Complexity of organization*: cells are composed of large numbers of molecules (usually more than 1 trillion in a typical cell). The molecules are organized into compartments, membranes and other structures in the cell. Bacteria are considered to have the simplest cells known, yet such a cell contains at least 600 different kinds of proteins and 100 other substances. Other organisms are more complex.
- *Adaption to the environment*: living organisms respond to their environment, to air, light, water, soil, etc. Natural selection leads to adaption to their environment. Today, many species are threatened with extinction because they are not able to adapt fast enough to the changing environment.

2.1.3 Evolution of Life

Nature has always worked bottom up, from the simplest assemblies to more complex structures. There exist many examples of self-organization in biology: spontaneous folding of proteins and other biomacromolecules, formation of lipid bilayer membranes but also social animals like social insects (bees, ants, termites), formation of flocks by birds etc. Molecular self-assembly is crucial to the function of cells: self-assembly of lipids to form the membrane, the formation of double helical DNA through hydrogen bonding of the individual strands etc. Molecular self assembly seems to be the key for the transition from non-living to living substances. The question is, however, what are the conditions of the environment (temperature, pressure, atmosphere, solvents) that are required that such a self assembly occurs.

In the 1950s it was assumed that the primitive Earth atmosphere consisted of methane CH_4, ammonia NH_3, hydrogen H_2 and water H_2O and S.L. Miller and H.C. Urey[1] carried out a famous experiment at the University of Chicago. They simulated the primitive Earth atmosphere and ran continuous electric currents simulating lightning storms, which were very common on the early Earth, to this environment. After one week, 10–15 amino acids were found in this primordial soup.

[1] In: Miller, S.L. "Production of amino acids under possible primitive Earth conditions", Science, 117, 528 (1953) and Miller, S.L., and Urey, H.C. "Organic compound synthesis on the primitive Earth", Science, 130, 245 (1959).

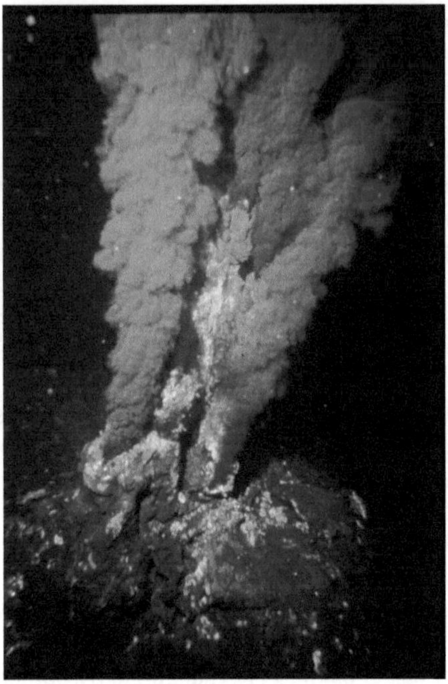

Fig. 2.2 Black smoker at a mid-ocean ridge hydrothermal vent. Credit: OAR/National Undersea Research Program (NURP); NOAA

Sagan and Chyba (1997) [287] proposed that the early Earth had an organic haze layer in its atmosphere. Such a layer can be found in the atmosphere of Titan, the largest satellite of Saturn and is produced by methane photolysis in the presence of nitrogen. An organic haze layer would preferentially absorb ultraviolet light, thereby allowing ammonia and methane to persist in the atmosphere.

In the 1970s black smokers (Fig. 2.2) were detected. These are chimney-like structures above hydrothermal vents.[2] In these smoker chimneys sulfides of iron, copper and zinc are found. At the mixture of the hot mineral rich water with cold water, these sulfides are precipitated and the vent water therefore appears black in color.[3] The most striking discovery was that these warm chemical rich environments are the living space for many species. Huber and Wächtershäuser, 1998 [166], modelled volcanic or hydrothermal settings. They showed that amino acids were converted into their peptides by use of precipitated (Ni, Fe)S and CO in conjunction with H_2S (or CH_3SH) as a catalyst and condensation agent at $100^\circ C$ and pH 7 to 10 under anaerobic, aqueous conditions. Thus a thermophilic origin of life seems plausible.

Summarizing there are basically three different theories about where life has evolved on our planet: the primordial soup theory (Urey-Miller experiment), life has originated in the atmosphere (Sagan) and black smokers (hydrothermal vents). The primordial soup theory seems to be no longer acceptable without strong mod-

[2]Sometimes also called underwater geyser.

[3]Hydrogen sulfide H_2S is the closest chemical analogue of water.

ifications since the primitive Earth's atmosphere contained much less reducing molecules. It did not consist mainly of NH_3, CH_4 but most probably of CO_2, CO, N_2. In any case water played an important role by acting as a solvent for the different molecules.

In the early twentieth century S. Arrheniusdeveloped his panspermia hypothesis. According to this theory life might have originated somewhere in the universe and would have spread out automatically.

The origin of life seems to be strongly connected to the presence of water for another reason. The Sun emits also UV radiation which is extremely hostile to life. Today we are protected against this radiation by the Ozone layer. This layer was slowly formed when the plants enriched the atmosphere with free oxygen. At the time when the first cells developed (about 3.8 billion years ago), there was no UV protection because there was no free oxygen in our atmosphere. However water, liquid or as ice, provides a good protection against this radiation. Therefore, it seems logical that life originated in such an environment.

Because the early Sun was less luminous, the temperatures could have been lower on Earth. A global frozen ocean (several 100 m thick) could have provided an ideal shield against UV radiation.

A frozen Earth could also be the result of a close encounter of the solar system with a nearby passing star, when the Earth is expelled out of the solar system or at least to larger distances from the Sun. Adams and Laughlin, 1999 [3] showed that for some time due to radiogenic heating life would still be possible around hydrothermal vents. Nisbet and Sleep, 2001 [248], discuss early hyperthermophile life near hydrothermal systems. The development of anoxygenic and then oxygenic photosynthesis would have allowed life to escape the hydrothermal setting. By about 3 500 million years ago, most of the principal biochemical pathways that sustain the modern biosphere had evolved.

The construction of the basic building blocks for life (monomers) is easy to explain (extraterrestrial origin, Urey-Miller experiment, black smokers...). However, it is much more difficult to explain the formation of polymers out of monomers. According to Darwin's Soup theory, mixtures of monomers should produce polymers. For the process of polymer production water plays also a negative role in an aequous environment, hydrolysis transforms the polymers into their constituent monomers. By hydrolysis the water molecule H_2O is split into H^+ and OH^-.

In this context we also mention autocatalytic reactions. In such reactions, one or more of the products are the same as one or more of the reactants. For example

$$A + B \rightleftharpoons 2B \tag{2.1}$$

The rate equations are non linear. The non linearity can lead to the spontaneous creation of order out of chaos. In nature there are many examples for such processes. Out of the random motion of air molecules a hurricane can be formed where the molecules all follow a vortex motion. This does not stand in contradiction with the second law of thermodynamics. The order created by living systems on Earth is produced at the expense of the increasing entropy inside the Sun which provides the necessary energy.[4] In 1995 Stuart Kauffman proposed that life initially arose as autocatalytic chemical networks.[5]

2.1.4 *Life Under Extreme Conditions*

In many textbooks you will find that for life to exist, the presence of liquid water is a necessary condition. Even habitability is defined very often as the zone around a star in which, on a hypothetical planet, water can exist in liquid form.

However, there are organisms that live under extreme physical or geochemical conditions, the extremophiles. Most of them are microbes including representatives of all three domains (Bacteria, Archaea, and Eucarya). Some types of extremophiles are listed in Table 2.1.

As it can be seen from the table, there are organisms that live under extreme "dry" conditions: the endoliths live in microscopic space within rocks, halophiles need salt to grow, the endolithic or cryptoendolithic microorganisms thrive inside rocks or in the pores between mineral grains of a rock.

The hyperthermophiles were discovered in the 1960s in hot springs in Yellowstone National Park. The hyperthermophile *Strain 121* is able to survive a temperature of 121°C and even to double its population within 24 hours in such an environment. It consists of hyperthermostable proteins. Another example is *Pyrolobus fumarii*, an Archaea living at 113°C in Atlantic hydrothermal vents.

Autotrophs produce food from inorganic compounds, the heterotrophs use organic molecules as food. Some bacteria and some archaea have this ability.

But note that all these lifeforms consist of cells with cytoplasm, where water is an important part. The cytoplasm is the part of a cell that is enclosed within the plasma membrane. For example halophilic bacteria produce energy to exclude salt from their cytoplasm to avoid protein aggregation.

The detection of life under such extreme conditions makes us hope to find life also under the harsh conditions on our neighbor planet Mars.

Cavicchioli, 2002 [64] provided a review on extremophiles and the search for extraterrestrial life.

2.2 Water and Other Solvents

2.2.1 *The Importance of Solvents to Life*

Solvents are extremely important for the formation and maintaining of life. Their main properties are:

---

[4]See also: Ilya Prigogine (1980). From Being to Becoming: Time and Complexity in the Physical Sciences. San Francisco: Freeman.

[5]Stuart Kauffman (1995). At Home in the Universe: The Search for the Laws of Self-Organization and Complexity. Oxford University Press.

Table 2.1 Some types of extremophiles

Name	Environment
Acidophiles	Live in pH levels ≤ 3
Alkaliphiles	Live in pH ≥ 9
Barophiles	Live under high pressure
Piezophiles	–
Endoliths	Live in microscopic spaces within rocks
Cryptoendoliths	Live in fissures, faults in deep subsurface
Halophiles	Need salt to grow
Hyperthermophiles	Live at $T = 80$–$120°C$, hydrothermal vents
Hypoliths	Live underneath rocks in cold deserts
Litoautotrophs	Source of carbon is CO_2, derive energy from mineral compounds
Metallotolerants	Capable of tolerating high levels of metals like copper, cadmium, zinc
Oligotrophs	Grow in nutritionally limited environment
Osmophiles	Grow in sugar concentration
Piezophiles	Live at high hydrostatic pressure; e.g. oceanic trenches
Polyextremophiles	Extremophiles more than one category
Psychrophiles, cryophiles	Grow at $T < -15°$; permafrost, polar ice...
Radioresistants	Tolerate high levels of ionizing radiation
Thermophiles	Live at temperature up to 80°C
Thermoacidophiles	Prefer temperatures of 70–80°C and pH between 2 and 3
Xerophiles	Grow in extremely dry conditions

- dissolution of chemical compounds,
- transport of nutrients in a cell,
- transport of waste,
- mixing of different chemical compounds,
- regulation of temperature; a constant temperature is required for complex organisms.

To be useful, any solvent must remain liquid within a large range of temperatures. Otherwise variations in conditions on a planet or satellite of a planet will freeze or boil the solvent and living organisms will be destroyed.

Under conditions like on the Earth's surface, the temperature range for water to remain liquid is 100°C. As the external pressure decreases (e.g. on a high mountain), the boiling point of water decreases (see Chap. 1).

For regulating the temperature, the heat capacity is the relevant parameter. Water has a high heat of vaporization; a living cell can respond to a temperature increase by vaporizing just a small amount of water. The heat capacity of water is 4.19 J/g/K,[6] the heat capacity of ammonia is 5.15 J/g/K and of methyl alcohol 2.15 J/g/K. To vaporize water requires 2491 J/g while ammonia needs 1256 J/g and methyl alcohol only 1214 J/g. Thus water seems to be ideal for temperature regulation. Mammals have a complex brain and precise temperature regulation allows to function this complex brain and cell systems properly.

In order to form aggregates of organic compounds, the surface tension of a solvent is relevant and water has a high surface tension. Thus water is much more appropriate for life regarding its heat capacity, heat of vaporization, surface tension and dissolving capability than other substances. Another important property of water is that it extends as it freezes. The cells of organisms that freeze will rupture. Putting the crew of an interstellar spaceship on ice would not help since humans consist mainly of water and the cells are destroyed by freezing. Ammonia does not expand upon freezing.

Another aspect has to be considered: as we have mentioned, stars produce UV radiation. Water molecules will be dissociated under intense UV radiation and the free oxygen contributes to ozone production. A planet with oceans consisting of ammonia does not produce such a shield.

2.2.2 *Other Solvents than Water*

In this section we will discuss two other well known solvents, ammonia and methyl alcohol.

The temperature range for liquid ammonia is from −78 to −33°C. The temperature range for liquid methyl alcohol is from −94 to +65°C. The parameter surface tension for ammonia and alcohol (ethanol, methanol) is only 1/3 that of water.

Life based on ammonia or methyl alcohol will be in some respects more robust and better protected against cosmic catastrophes than life based on water. However, it seems that complex life cannot be based on these because an effective regulation of temperature of the organisms is needed.

Let us speculate further. There is one advantage for Astronauts whose life chemistry depend on ammonia as solvent: they could easily be frozen and then waked up at arrival on some distant stellar system since ammonia does not expand when frozen and the cells are not destroyed.

---

[6] Also called 1 calorie.

2.3 Energy for Life

2.3.1 Energy

To keep biological processes running, a constant supply of energy (mainly from the Sun) is needed. The two laws of thermodynamics are fundamental: (i) Energy is conserved, neither created nor destroyed. Energy may be transformed e.g. from the energy in a chemical bond to heat energy; the total amount always remains unchanged. (ii) Entropy measures disorder in natural systems. The entropy tends to increase in all natural systems.

These two laws can be applied to biological systems. Organisms are highly organized. This organization can only be maintained by a constant supply of energy. This energy is used by the cell to do work, some of that energy is lost as heat, we call this dissipation.

2.3.2 Metabolic Diversity

Metabolism can be defined as by all chemical reactions that occur inside an organism.

ATP (Adenosine triphosphate) transports chemical energy within cells for metabolism. One molecule of ATP contains three phosphate groups, and it is produced from inorganic phosphate and adenosine diphosphate (ADP) or adenosine monophosphate (AMP). The standard amount of energy released from hydrolysis of ATP is:

$$\mathrm{ATP} + \mathrm{H_2O} \rightarrow \mathrm{ADP} + \mathrm{P}_i, \qquad \Delta G = -30.5\ \mathrm{kJ/mol} \tag{2.2}$$

$$\mathrm{ATP} + \mathrm{H_2O} \rightarrow \mathrm{AMP} + \mathrm{PP}_i, \qquad \Delta G = -46.5\ \mathrm{kJ/mol} \tag{2.3}$$

P denotes a phosphate, PP_i a pyrophosphate.

Prokaryotes[7]use many compounds to obtain energy in the form of ATP and this enables them to habitate a wide variety of environmental habitats. The process to obtain energy in non photosynthetic cells is as follows:

- energy source: this source provides an electron,
- this electron is transferred to,
- electron acceptor.

An oxidation is the following reaction:

$$A \rightarrow A^+ + e^- \tag{2.4}$$

These organisms are called chemotrophs. Organisms that derive energy from sunlight are called phototrophs. The chemotrophs are further divided into:

[7]These organisms have no cell nucleus. Most are unicellular; they are divided into bacteria and archaea.

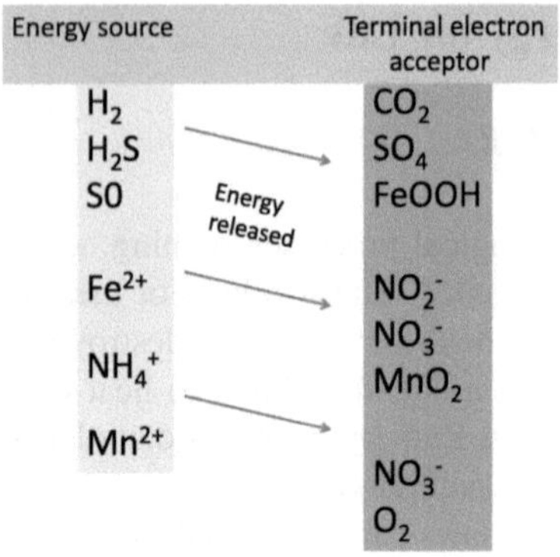

Fig. 2.3 Different types of chemotrophic metabolism

- chemoorganotrophs: they oxidize organic compounds;
- chemolithotrophs: obtain energy by oxidizing inorganic chemicals

In Fig. 2.3 different types of chemotrophic metabolism are given.

Let us consider for example the methanogens: they generate ATP by oxidizing hydrogen gas, using CO_2 as terminal electron acceptor.

$$4H_2 + CO_2 \rightarrow CH_4 + H_2O \tag{2.5}$$

Methanogens are found in anaerobic environments where the gases hydrogen and carbon dioxide are available. These are produced by chemoorganotrophs using fermentation of organic material.

2.3.3 *Solar Energy*

The Sun provides warmth. This is essential for all organisms most of which can survive only within a narrow temperature range. At high temperature (above 40°C), the biomolecules start to break down, at low temperatures (near 0°) the chemical reactions for metabolism occur too slowly. The organisms stop to grow and reproduce. Two important factors help to moderate and maintain temperatures on Earth: the earth's atmosphere and the oceans.

At the top of our atmosphere 1.372 kW/m^2 (1 W = 1 J/s) solar energy is received. More than half of the incoming sunlight may be reflected or absorbed by clouds, dust and gases. Short wavelength radiation is filtered out (e.g. UV by ozone) and cannot reach the surface.

The distribution of the incoming solar radiation is illustrated in Fig. 2.4. Incoming solar radiation is absorbed by land and sea and reflected into space by water, snow and also land surfaces. Half of the energy plants absorb is used in evaporating water, only 1–2% of the sunlight falling on the surface is available for photosynthesis. However, this small percentage is the energy base for any life in the biosphere.

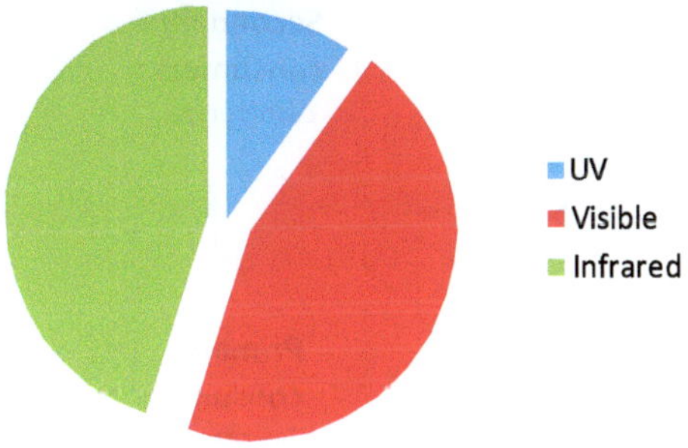

Fig. 2.4 The different percentages of solar radiation that reach the Earth's surface. Photosynthesis use *blue* and *red* light; most planets reflect *green* wavelengths, therefore, they appear *green*

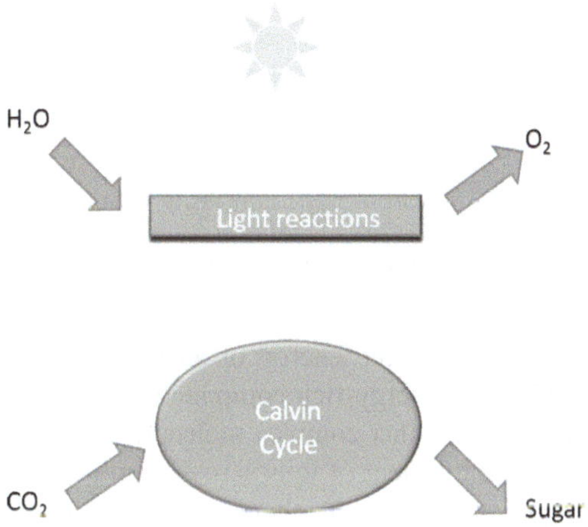

Fig. 2.5 A simplified diagram of photosynthesis

2.3.4 Photosynthesis and Respiration

Within plant cells there are the chloroplasts that contain chlorophyll. The chlorophyll molecules absorb light energy and create chemical bonds that serve as the fuel for all subsequent cellular metabolism. There occur two types of reactions:

- Light-dependent reactions: the chloroplast receives light. Enzymes split water molecules and release molecular oxygen, O_2. This is the source for all the oxygen in our atmosphere on which all animals depend for life. During the light-dependent reaction two types of mobile, high-energy molecules are created: ATP (adenosine triphosphate) and NADPH (nicotinamide adenine dinucelotide phosphate). These provide energy for the next process.
- Light-independent reactions: the enzymes extract energy from ATP and NADPH adding carbon atoms (from carbon dioxide) and the result is a sugar molecule (e.g. glucose). These glucose molecules provide the building blocks for larger, more complex organic compounds.

The photosynthetic reactions can be summarized as (see also Fig. 2.5):

$$6H_2O + 6CO_2 + h\nu \rightarrow C_6H_{12}O_6 + 6O_2 \tag{2.6}$$

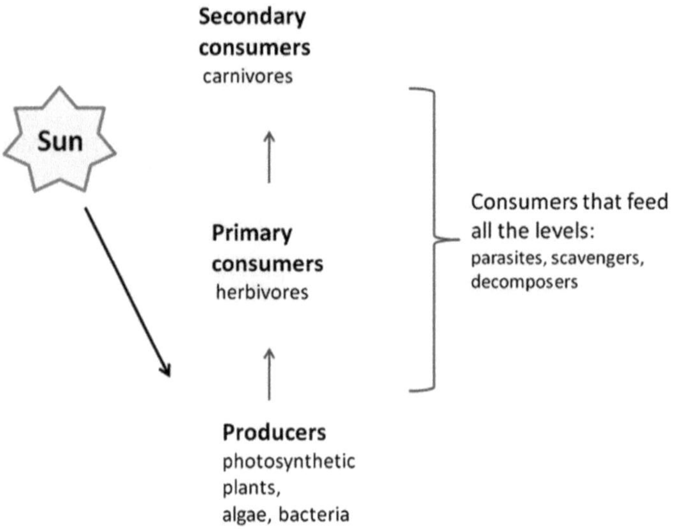

Fig. 2.6 How organisms obtain food for their life processes

where $h\nu$ denotes the energy of an incoming solar radiation quantum. Glucose, $C_6H_{12}O_6$ is an energy-rich compound. Other enzymes release the energy in these compounds and other complex molecules such as lipids, proteins, nucleic acids are formed, or movement of ions across membranes, changes in cellular structure etc. Photosynthesis occurs in plants, algae, and many species of Bacteria, but not in Archaea. The amount of energy trapped by photosynthesis is approximately 100 terawatts per year.[8]

The reverse process to photosynthesis is cellular respiration (Fig. 2.6 summarizes both the connection between both processes). From the sugar molecule, carbon dioxide and water is released.

$$C_6H_{12}O_6 + 6O_2 \rightarrow 6H_2O + 6CO_2 + \text{energy} \tag{2.7}$$

Note that:

- photosynthesis: energy is captured,
- respiration: energy is released.

Animals and humans eat plants or other animals that have eaten plants. Organic molecules in the food are broken through cellular respiration to obtain energy.

Note that in both, photosynthesis and respiration, water plays an important role. Photosynthesis as an energy source is limited to at most the top few hundred meters of water bodies, and to the land surface.

[8]This is about seven times larger than the yearly power consumption of human civilization.

Chapter 3
Water on Planets and Dwarf Planets

In this chapter the planets and dwarf planets of the solar system are discussed. First we give a short description about their most relevant physical properties and structure and then the occurrence of water in their atmospheres and surfaces is reviewed. Water and related chemistry in the solar system is a Herschel Space Observatory[1] Guaranteed-Time Key Programme (Hartogh et al. 2010 [158]). The European Space Agency, ESA, approved this program and the main goals are to study the evolution and origin of water in Mars, the outer planets, Titan, Enceladus and comets.

3.1 Classification of Objects in the Solar System

3.1.1 Overview

The objects in the solar system can be classified as follows:

- The Sun: it is our nearest star, at a distance of about 150×10^6 km which is also used as a distance unit referred to as astronomical unit, AU. The Sun contains 99.8% of the total mass of the solar system and is thus by far the dominating mass in it. All other objects in the solar system orbit the Sun.
- Planets: the eight great planets are Mercury, Venus, Earth, Mars, Jupiter, Saturn, Uranus and Neptune. Five of them can be seen with the naked eye and were known by the ancients (Mercury, Venus, Mars, Jupiter and Saturn).
- Dwarf Planets: since 2006 this category of objects was introduced to take into account the detection of several objects that are about the size of Pluto. Furthermore, their orbits have not been totally cleaned up. Their shape is almost spherical.
- Comets and Asteroids: only few asteroids are larger than 100 km, comets are even smaller.
- Small particles like meteoroids, interplanetary dust particles.

[1] Successfully launched in May, 2009.

A. Hanslmeier, *Water in the Universe,* Astrophysics and Space Science Library 368,
DOI 10.1007/978-90-481-9984-6_3, © Springer Science+Business Media B.V. 2011

Table 3.1 Some important parameters of the planets in the solar system. D denotes the distance from the Sun, P the orbital period, R the radius and P_{Rot} the rotation period

Planet	$D \times 10^6$ km	P	R (km)	P_{Rot}
Mercury	57.91	88.0 d	4879 (0.38)	58.65 d
Venus	108.21	224.7 d	12 103 (0.95)	−243.02 d
Earth	149.6	365.25 d	12 742 (1.0)	23 h 56 m
Mars	227.92	687.0 d	6780 (0.53)	24 h 37 m
Jupiter	778.57	11.75 y	139 822 (10.97)	9 h 55 m
Saturn	1433.53	29.5 y	116 464 (9.14)	10 h 40 m
Uranus	2872.46	84 y	50 724 (3.98)	−17.24 h
Neptune	4495.06	165 y	49 248 (3.87)	16.11 h

Asteroids, comets and smaller particles are also grouped into SSSBs, small solar system bodies. All categories of objects described above appear at specific locations in the solar system. The inner solar system contains the terrestrial planets and the Main Belt of asteroids. In the middle region there are the giant planets with their satellites and the centaurs.[2] The outer solar system comprises the Trans-Neptunian objects including the Kuiper Belt, the Oort cloud, and the vast region in between.

The solar system is not unique in the universe, in the beginning of 2010 about 400 extrasolar planetary systems were known. The formation of such systems seems to be a normal process.

3.1.2 *Physical Parameters of Planets*

Up to now all planets of the solar system have been explored also by satellite missions and on Mars and Venus even some successful unmanned landing missions were made supplying us with really in situ surface measurements.

Some important data of the planets are given in Table 3.1. In this table D denotes the distance from the Sun, P the period of revolution about the Sun, R the radius of the planet and P_{Rot} the rotation period. The minus sign in the column for the rotation period indicates rotation in the sense opposite to the revolution about the Sun.

3.2 Terrestrial Planets

Terrestrial planets have a solid surface, are differentiated (lighter elements near the surface, heavier elements in the core) and have relatively high densities. The main spectral signatures between 8 and 20 μm in the atmosphere of Venus Earth and Mars are shown in Fig. 3.1.

[2] A group of asteroids that cross the orbits of the giant planets.

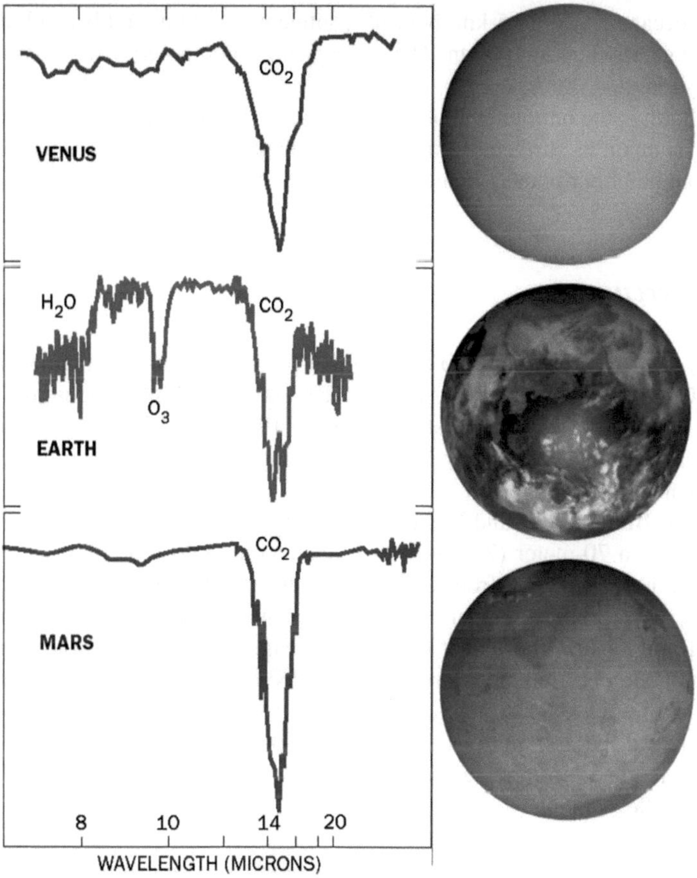

Fig. 3.1 The main spectral signatures coming from the atmosphere of the terrestrial planets. Note the strong carbon dioxide absorption in the spectra of Venus and Mars. Source: www.markelowitz.com/exobiology.htm

3.2.1 Earth

We have already discussed where there is water and how it is distributed on Earth. Our Earth is the largest of the terrestrial planets. The mass is 5.97×10^{24} kg, the density $5.5\ \mathrm{g\,cm^{-3}}$. The oceans comprise 2/3 of its surface. Over the whole evolution, the mass of land has increased steadily, during the past two billion years, the total area of the continents has doubled. Like the other terrestrial planets, the interior of the Earth can be divided into layers: the lithosphere is found at a depth between 0–60 km, the crust between 5 and 70 km, the mantle between 35 and 2890 km, the outer core between 2890 and 5100 km and the inner core between 5100 and 6371 km (density $\sim 12\ \mathrm{g\,cm^{-3}}$). The geothermal gradient is the rate of increase in temperature per unit depth in the Earth. This internal heat is produced by radioactive decay of ^{40}K, ^{238}U, ^{232}Th. The rigid crust is very thin compared with the other layers, beneath oceans it is only 5 km, beneath continents ~30 km and beneath mountain ranges (e.g. Alps) up to 100 km. The atmosphere is composed of 78% nitrogen and 21% oxygen.

The Earth is the only known object in the solar system where water can be found in liquid state on its surface. If this is a prerequisite for life, then the chances to find life on our neighbor planets Venus and Mars are very low.

3.2.2 Mercury

Mercury is the closest planet to the Sun and the temperature on the dayside is up to 450°C and on the nightside −180°C. This extremely high temperature contrast is a consequence of lacking an atmosphere. In 1974 and 1975, the US-Mariner 10 spacecraft made three passes by Mercury, sending back photos covering 45 percent of the planet's surface.

In 1991, Butler, Slade and Muhleman [53] studied Mercury using a radar system consisting of a 70-meter (230-foot) dish antenna at Goldstone, CA. The beam of 8.5-GHz microwaves sent from Goldstone bounced off Mercury and was collected at the VLA[3] to produce a radar image of the planet. They found some strong radar reflecting regions on the planet's surface. In Fig. 3.2 the bright dot at the top of the disk of Mercury indicates strong radar reflection at Mercury's north pole. This reflection resembles the strong radar echo seen from the ice-rich polar caps of Mars.

This led to the assumptions that ice could persist inside deep craters near the poles of Mercury. Normal ice absorbs radar signals, but ice at extremely low temperatures reflects them.

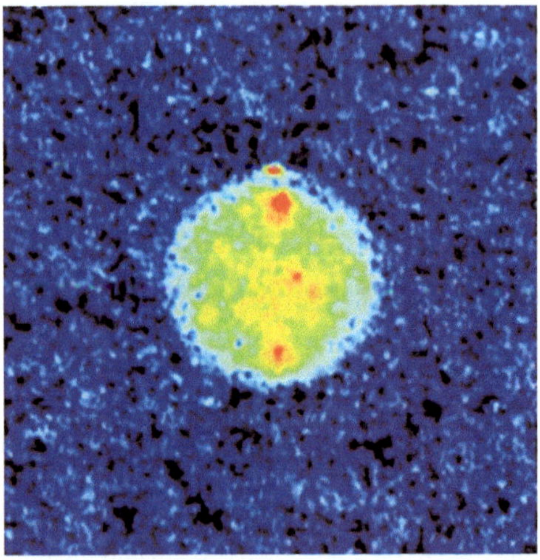

Fig. 3.2 Radar image of Mercury; *red* indicates strong reflection of the radar signal and *yellow*, *green*, and *blue*, progressively weaker reflection. The bright *red dot* at the top of the image indicates strong radar reflection at Mercury's north pole. Image Credit: NRAO, AUI, NSF

[3]The Very Large Array consists of 27 radio antennas in a Y-shaped configuration fifty miles west of Socorro, New Mexico. Each antenna is 25 meters in diameter.

Temperatures in the polar regions of Mercury and implications for water ice were discussed by Wood, Vasavada, and Paige, 1992 [363]. Radar observations of Mercury have revealed the presence of anomalous radar reflectivity and polarization features near its north and south poles where it never gets warmer than 170 K. Therefore it is argued that maximum surface temperatures in shaded cratered regions near the poles are below 110 K and at a temperature lower than 112 K water ice could be stable to evaporation for several billion years (see also Paige et al., 1992 [258]).

Where does this ice on Mercury come from? Exogenic sources of water for Mercury's polar ice are discussed by Rawlins, Moses, and Zahnle, 1995 [276]. Continual micrometeoritic bombardment of Mercury over the last 3.5 billion years could have resulted in the delivery of (3–60) $\times 10^{13}$ kg of water ice to the permanently shaded regions at Mercury's poles (equivalent to an average ice thickness of 0.8–20 m) as has been estimated by Moses et al., 1999 [237].[4]

It was shown by Starukhina, 2002 [324] that cold silicates could also produce high radar reflection and thus the radar brightness of the polar craters of Mercury does not necessarily need the existence of water ice there.

MESSENGER (Mercury Surface, Space, ENvironment, GEochemistry, and Rangin) is a very ambitious mission to Mercury. The spacecraft was launched on August 4, 2004 and performed two Venus flybys in 2006 and 2007 and made its first Mercury flyby on Jan 14, 2008. Mercury orbit insertion is planned in 2011. Head et al., 2009 [160] reported on widespread nature of volcanism, the presence of pyroclastic deposits, and the volcanic filling of impact craters and basins on Mercury.

3.2.3 Venus

Venus is sometimes referred to as "sister Earth" because it is similar in size. Its surface is hidden by dense clouds (Fig. 3.3). Observations made from ground by radar and from satellite missions and even successful landings revealed an extremely dry surface resembling the surface of Mercury or the Moon, so the conditions for life are not very promising. The surface pressure of the atmosphere is 90 times that of Earth's,[5] and, because of the dense atmosphere that consists mainly of the greenhouse gas CO_2, the surface temperature is about 460°C. It is interesting to note, that this high surface temperature can be found on any place on the surface of Venus, there are no big differences between e.g. the equator and the poles. Therefore, contrary to Mercury, it is unlikely to find ice or water near the poles of Venus. Venus rotates in 243 Earth days in the direction opposite to its orbital motion and one revolution about the Sun lasts 224.7 days. The surface of Venus shows craters and ancient volcanoes (Fig. 3.4).

---

[4] The Antarctic ice sheet is about 30 million km^3, an amount equivalent to 70 m of water in the world's oceans.

[5] That corresponds to the pressure at a water depth of 900 m on Earth.

Fig. 3.3 Ultraviolet image of Venus' clouds as seen by the Pioneer Venus Orbiter (February 26, 1979). NSSDC, NASA

Fig. 3.4 Venus: Meat Mons, an 8 km high volcano. Magellan synthetic aperture radar data is combined with radar altimetry to develop a three-dimensional map of the surface. NASA, Magellan Mission

In the atmosphere of Venus there occur thick H_2SO_4 clouds extending from 40 to 60 km altitudes. The region from 60 to 100 km is called mesosphere. At the top of the thick clouds there is a four day superrotation, above 100 km there is a solar-antisolar circulation. In the mesosphere the most important trace gases are CO, SO_2, HCl, HF, H_2O and HDO.

Water Vapor

Water vapor is an important trace gas in the lower Venus atmosphere and it contributes to the global greenhouse effect that maintains the high surface temperature. Most of the H_2O is found below the cloud base at about 47 km. Its signature can be observed in a spectral window between 0.9 and 2.5 μm where the atmospheric absorption is weak, therefore H_2O absorption features can be seen. For H_2O abundance determinations the following windows are most often used: 1.18, 1.74 and 2.3 μm.

Let us compare the water content in the atmosphere of Venus with that in the terrestrial atmosphere:

- Earth: average surface temperature 288 K, water vapor content in the troposphere up to 4% (Lodders and Fegley, 1998 [207]).
- Venus: surface temperature range from 660 K (on the high Maxwell mountains) to 740 K (in the plains), the water content in the subcloud atmosphere is about 30 ppmv[6] (Bailey, 2009 [15]).

Another technique to measure the water content in a planetary atmosphere is using solar occultation in the IR. Such experiments can be made during sunrise or sunset. Before the instrument's line of sight to the Sun intersects with the top layers of the atmosphere, the measurement cycle is started, and as soon as the top of the atmosphere is reached, solar light is absorbed and the intensity of the recorded signal starts to decrease. When molecules start to absorb the radiation, structures appear in the spectrum. Those structures are characteristic of a specific molecule and their amplitude or depth are in direct relation to the quantity of this species present in the sounded atmosphere.

Vertical distributions of the molecular density and mixing ratios of H_2O and HDO in the Venus mesosphere were measured by Fedorova et al., 2008 [129]. The experiment was carried out on board of Venus express mission (SOIR instrument, 2.32–4.35 μm). The atmosphere was sounded during solar occultation in the range of altitudes from 65 to 130 km. An enrichment of D to hydrogen indicates the escape of water from Venus. Bertaux et al., 2007 [25] report on the detection of a warm layer at 90–120 km.[7]

Atmospheric Escape

The Venus Express mission was launched in November 2005 and is on a quasi polar orbit around Venus. The mission operations will be extended up to 2012. One of the main tasks of Venus Express is to answer the question where the water on Venus has gone. For the first time it was possible to measure directly the water loss of Venus which is about 10^{24} molecules per second (Delva et al., 2009 [105]). Using

---

[6]Parts per million by volume.

[7]For altitudes larger than 100 km a cold cryosphere is assumed on the nightside of Venus.

the Analyser of Space Plasma and Energetic Atoms (ASPERA) on board Venus Express revealed a great loss of hydrogen and oxygen on the night-side. Roughly twice as many hydrogen atoms as oxygen atoms were escaping. Because water is made of two hydrogen atoms and one oxygen atom, the observed escape indicates that water is being broken up in the atmosphere of Venus. This water escape can be explained by the fact that Venus does not have a protective magnetic field like the Earth. Therefore, the energetic solar wind particles strike the upper atmosphere and carry off particles into space. Thus the interaction between the unprotected Venus atmosphere and the solar wind can explain the present hydrogen and oxygen loss.

Atmospheric and water loss from early Venus was studied by Kulikov et al., 2006 [190]. Pioneer Venus mass spectrometer data of the deuterium to hydrogen (D/H) ratio of 120 ± 40 times the terrestrial value indicate that Venus may have had at least an H_2O content of the order of about 0.3% of a terrestrial ocean, and even much more during and shortly after the accretion period during the first 300 Myr of the solar system history. During the accretion phase of the solar system, the protosun was already formed and tiny particles started to grow rapidly to form planetesimals which then collided to form the planets. In the inner part of the solar system metals and rock condensed, in the outer parts ice. After the accretion phase, an additional unknown amount of H_2O was supplied to Venus by cometary impacts. During its early evolution, Venus has lost most of its H_2O because of the very active young Sun: high X-ray, EUV and solar wind flux.

Aboard the Venus Express spacecraft the Visible and Infrared Thermal Imaging Spectrometer (VIRTIS) constructed a map of water vapor abundance in the lower atmosphere by nighttime observations in the 1.18 μm band. The thermal emission from this window orignates from heights 0–15 km. Practically no latitudinal variations were found, the water vapor mole fraction is about 44 ± 9 ppm (Bézard et al., 2009 [29]). A distribution modeling of H_2O and HDO in the upper atmosphere of Venus was presented by Liang and Yung, 2009 [202].

The evolution of Venus Earth and Mars will be compared in a separate chapter.

3.2.4 Mars

Mars is the second closest in similarity to Earth. Due to the turbulent Earth's atmosphere, it is difficult to observe details on its surface from earthbound observatories. Moreover, because of its elliptic orbit, the closest distance to Earth can range from about 56 Million to more than 100 Million km. Turbulence in the Earth's atmosphere blurs and distorts the image (this effect is known as seeing in astronomy) and can lead to completely wrong interpretation of observations such as the canali on Mars. Schiaparelli observed Mars during its approach to Earth in 1877. He claimed to have found on the surface of Mars a network of channels which he called canali. Later, it was suggested by Lowell and Flammarion that these canali were constructed by intelligent Martians to distribute water on the dry planet. From that time on (beginning of the twentieth century) the legend of little green men on Mars was born.

Fig. 3.5 Mars seen from the Earth orbiting Hubble Space Telescope, HST. One of the polar caps is clearly seen as well as some surface features and clouds in its atmosphere. Image Credit: HST

Many attempts were made to investigate the red planet with satellite missions, however, until 2006 only 18 of 37 launch attempts to reach Mars have been successful.

The first images of the surface of Mars from a satellite mission were obtained in 1965 by Mariner 4 (US). In 1971 for the first time a satellite could be brought into orbit around a planet (with exception of Earth of course). The results were disappointing because the surface of Mars appears more like that of the Moon and there are no signs of the canali claimed to have been observed.

Mars, Surface

A high resolution image of Mars taken from the Hubble Space Telescope is shown in Fig. 3.5. On a nice summer day, the temperature on Mars may rise up to 0°C however during a Martian night it may reach −100°C. There are channel like structures[8] on its surface which are a hint that this planet underwent large climatic variations in the past with episodes of liquid water on its surface. Today, because of the low atmospheric pressure (only 1% of the pressure on the surface of the Earth) water cannot exist in liquid state on Mars. For liquid water on the martian surface, the atmosphere must become much denser (see also Chap. 1). Water can only sublimate on the surface of Mars i.e. it undergoes a phase transition from solid (ice) to gas (vapor) or from gas to solid.

[8]These channels are far too small to have been detected from telescopic observations from Earth.

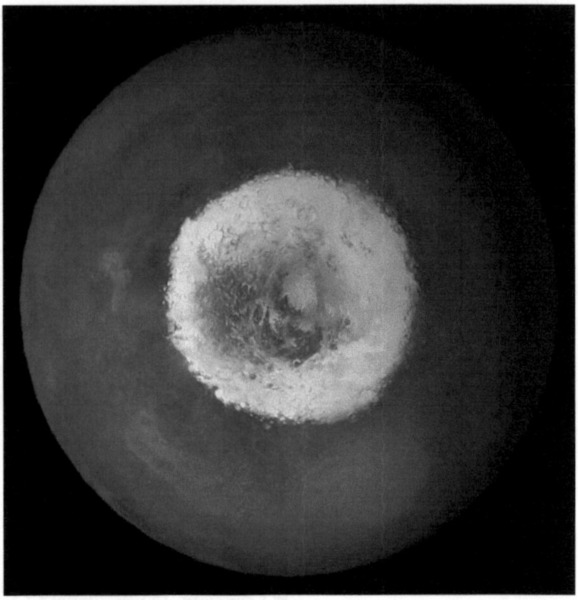

Fig. 3.6 South polar region of Mars. Image Credit: NASA/JPL/Malin Space Science Systems

Two polar caps exist on Mars (see Fig. 3.6). The permanent portion of the north polar cap consists almost entirely of water ice. In the northern hemisphere winter an additional coating of frozen carbon dioxide about one meter thick is deposited. The south polar cap also acquires a thin frozen carbon dioxide coating in the southern hemisphere winter. Beneath this is the perennial south polar cap, which is in two layers. The top layer consists of frozen carbon dioxide and about 8 meters thick. The bottom layer is very much deeper and is made of water ice. In 2005 data from NASA's Mars Global Surveyor and Odyssey missions revealed that the carbon dioxide ice caps near Mars's south pole had been diminishing for three summers in a row. This can be regarded as a sign of long-term increase in solar irradiance that affects both the Earth and Mars. Also albedo variations on Mars may play a role (Fenton, Geissler and Haberle, 2006 [131]).

Geomorphologic evidence for liquid water on Mars indicating a hydrologic cycle in the past was mentioned by Masson et al., 2001 [218]. These follow form the many channels and canyons observed on the surface of Mars.

High-resolution images from the Mars Obiter Camera (MOC) onboard the Mars Global Surveyor (MGS) show a variety of gully features on sloped surfaces of Mars. The mechanism of gully formation is still under debate. There are stronger water signatures (water absorption bands) at the gully-exposed sites than in the surrounding areas. This implies that the water signatures, most likely representing water ice, isolated water molecules, and/or hydroxyl molecules incorporated into minerals, are still present in the shallow unconsolidated soils. Therefore the formation of the gullies can be explained by water and gully formation is still locally active on the Martian surface in the present time (Fan et al., 2009 [125]).

More about methods of detection of water on Mars can be found in Chap. 9 of this book.

Water Vapor in the Atmosphere of Mars

Many measurements were made to determine the content of water vapor in the atmosphere of Mars. Some examples are the MAWD (Mars Atmospheric Water Detector) on board of Viking 1 and Viking 2[9] orbiters or measurements made with Mars Global Surveyor Mission[10] and Mars Express.[11] Very often the IR band about 1.38 μm was used. Such measurements are important input factors for understanding the annual water cycle on Mars. For example Fedorova et al., 2010 [128] report on Viking observation of water vapor on Mars explaining some discrepancies of the results obtained by different instruments.

Measurements of Martian water column abundances obtained by ground-based high-resolving-power spectroscopy in the very near IR (817.4–820.0 nm) from the 1.5-m Catalina observatory near Tucson, AZ were reported by Sprague et al., 1996 [319] for the period 1988–1995 and in Sprague et al., 2001 [320] measurements for the interval 1998–1999 were given. For example a strong increase in water vapor abundance from 8 to 76 ppt μm (precipitation micrometer) with onset of spring and early summer followed by an abrupt drop to less than 1 ppt μm by summer's end was found. Also the influence of long lasting dust storms on the water content can be seen.

3.2.5 *The Early Sun and Evolution of Terrestrial Planets*

Before discussing the occurrence of water on surfaces of the terrestrial planets at present and in the past we shortly describe the early Sun evolution since this had important influence on the water history of the planets.

The Sun formed by a collapse of an interstellar dust and gas cloud about 4.6 billion years ago. The formation of the Sun was a fast process, it took only several 10^6 years to reach the main sequence in the Hertzsprung-Russell diagram (see Chap. 9) where the temperature of a star is plotted versus its radiation. A main sequence star is characterized by hydrostatic equilibrium throughout its interior. Hydrostatic equilibrium is reached when compression due to gravity is balanced by a pressure gradient which creates a pressure gradient force in the opposite direction. Depending on its mass, a star may remain on this main sequence (see Chap. 9) for several billion years (for the Sun: about 9 billion years).

However, the early sun was quite different from the sun we know. Its luminosity was only about 70% of its present value. It rotated much faster. Stellar rotation is an important parameter to trigger stellar activity. The present day sun shows an activity cycle with a period of about 11 years. The number of sunspots varies but also the number of flare occurrence (these are energetic outbursts caused by a reconnection

---

[9]1974, 1975.

[10]Launched 1996.

[11]Launched 2003.

of magnetic field lines) and coronal mass ejections, CMEs. During such outbursts, energetic particles and radiation (UV, X-ray) are emitted. The solar wind streams off of the Sun in all directions at speeds of about 400 km/s. The source of the solar wind is the Sun's hot corona which is the outermost layer of the solar atmosphere with temperatures of several Million K. The solar wind speed is high (800 km/s) over coronal holes and low (300 km/s) over streamers and high and low speed streams interact with each other and alternately pass by the Earth as the Sun rotates. Such interactions can produce storms in the magnetosphere of the Earth. The solar wind consists mostly of electrons and protons with energies of about 1 keV. Due to the high temperature in the solar corona and due to some acceleration process these particles can escape from the gravity field of the Sun.

The early sun was less luminous but the energetic outbursts were of higher amplitude and occurred more frequent than presently. The UV emission was about 10, the extreme UV 100 and the X-ray emission 1000 times the present values. Interactions of high-energy radiation and the solar wind with upper planetary atmospheres may have led to the escape of important amounts of atmospheric constituents (Güdel, 2007 [148]).

The lower luminosity of the early Sun would imply lower surface temperatures on the terrestrial planets and e.g. the Earth would have been frozen totally. From geology we know that Earth was never frozen totally. This paradox is called the faint young Sun problem.

Lammer et al., 2008 [192] discussed the origin and evolution of Venus', Earth's, Mars' and Titan's atmospheres from the time when the active young Sun arrived at the Zero-Age-Main-Sequence. Thermal and various nonthermal atmospheric escape processes influenced the evolution and isotope fractionation of the atmospheres and water inventories of the terrestrial planets efficiently.

Several solutions to this problem have been suggested such as a high concentration of greenhouse gases in the early earth's atmosphere (see also Hanslmeier, 2007 [155]). Another suggestion to the early faint Sun paradox was made by Shaviv, 2003 [306]. Cosmic rays are supposed to have a cooling effect on the global climate on Earth. The stronger the flux of cosmic rays, the stronger the cooling. This could also explain the global temperature decrease during the Maunder Minimum which lasted from 1645–1715, a period where no solar activity was recorded. High solar activity means a high solar wind flux and a strong heliosphere that protects the planetary system from cosmic ray particles, hence less cosmic ray particles penetrate to the solar system during phases of high solar activity. The faint young Sun was much more active, therefore a stronger solar wind flux could be assumed. Consequently, during this faint young Sun evolution, the flux of energetic cosmic ray particles was lower than today, therefore, the climate was warmer on Earth. The relation between the temperature change due to cosmic ray flux ΔT_{CRF}, and the energy flux reaching the troposphere E is approximated by:

$$\Delta T_{\mathrm{CRF}} = D(1 - E/E_0)^q \tag{3.1}$$

where q is about $1/2$ and E_0 is the present day flux. The parameter D is in the range $10 \pm 5°\mathrm{C}$.

The expected temperature increase on Earth in the future because of the increasing solar luminosity is partly compensated by an increasing cosmic ray flux due to lower solar activity.

3.2.6 Dry Venus–Humid Earth–Climate Changes on Mars

After having discussed the terrestrial planets we address to the question why their evolution was so different concerning their water content. One important criterion is of course the surface temperature. On Mercury, the temperatures are high because of its close position to the Sun and there cannot exist water besides, possibly, near the poles.

It is reasonable to assume that Venus and Earth and even Mars accumulated similar amounts of primordial water during their accretion phase. So why did Venus lose its water?

Venus

The distance of Venus from the Sun is about 2/3 the distance of the Earth's. Theoretically, Venus could have quite favorable conditions for liquid water to exist on its surface. But there are two effects to consider. The present dense atmosphere of Venus with the greenhouse gas carbon dioxide causes too high surface temperatures. However, these unfavorable conditions might have been different in the early evolution of Venus. The luminosity of the early Sun was only about 70% of its present value. Therefore, on early Venus, water oceans could have existed. But the Sun became more luminous and due to its closer distance compared with Earth, Venus received more radiation from the Sun. Especially, the UV part of radiation has to be mentioned here. It causes a splitting of the water molecules which led to a runaway greenhouse effect. The lighter hydrogen escaped from the atmosphere, the heavier oxygen was bound with surface minerals. Such a scenario was described by Kasting, 1988 [181]. The other question concerns the evolution of Venus's atmosphere. This has to be seen in connection with plate tectonics. On Earth, gases like carbon dioxide are released but they are also absorbed by its active plate tectonics. On Venus, there are no signs of plate tectonics in recent time so its atmosphere became enriched with that greenhouse gas.

The problem of a wet early Venus was studied by several authors. Pioneer Venus mass spectrometer data suggest that the ratio of deuterium to hydrogen, D/H ration is 1.9×10^{-2}. This is about 120 ± 40 times the value found for the Earth (Lodders and Fegley, 1998 [207]). During evaporation, the lighter hydrogen preferentially escapes leaving the heavier deuterium. Donahue et al., 1982 [107], 1997 [108] concluded that Venus once must have had the equivalent of an at least 4 m thick global ocean (the maximum depth however could have been more than 500 m). It has to be stressed that the D/H value alone could be problematic in order to estimate the amount of water because there are examples of interplanetary dust particles and comets with high D/H ratios.

Therefore, additional proxies for water content on a planet must be investigated. A study of the surface mineralogy of Venus could provide another proof for the existence of large amounts of water during its early evolution phase. Under the presence of water, hydrous minerals formed on or below its surface. However, these minerals are unstable under the extreme temperature and pressure conditions on the surface (Zolotov et al., 1997 [376]).

Johnson and Fegley, 2000 [172], argued that the mineral tremolite might indicate a higher water content on early Venus. On Earth, tremolite is found in basic rocks in contact with siliceous carbonate rocks. One possible reaction for the formation is:

$$5CaMg(CO_3)_2 + 8SiO_2 + H_2O \rightarrow Ca_2Mg_5Si_8O_{22}(OH)_2 + CaCO_3 + 7CO_2 \quad (3.2)$$

The first two minerals are dolomite and quartz, on the right hand side the first is tremolite and the second is calcite. Tremolite can survive decomposition on Venus over geologic time scales at current and at higher surface temperatures.

At present there exist no mineralogical data of the Venusian surface. Abundances of Mg and heavier elements were measured by spectrometers on Venera 13 and Venera 14 and Vega 2 landers. Unlike the Moon and Mars, there are no known meteorite samples from the Venusian surface. Thus, if future spacecraft can detect tremolite, this discovery could serve as ultimate evidence of a wetter Venusian history.

The detailed loss of hydrogen by splitting up of water molecules in connection with less solar luminosity but higher amplitudes of activity (up to a factor of 100 in X-ray domain) was studied by Kulikov et al., 2006 [190]. The H_2O content of early Venus can be estimated to about 0.3% of the terrestrial ocean (Donahue et al., 1982 [107]). However, during the first 300 million years of the evolution of Venus, there might have been even much more water because of the intense bombardment phase—the number of impacts of cometary like bodies which mainly consist of water was considerably higher than today. The solar luminosity was less and therefore the hydrogen escape occurred also at a lower rate. During the active phase of the young Sun, the exospheric temperatures were considerably higher than today (about 8000 K) which lead to enhanced hydrogen diffusion. Therefore, the oxygen related to water that was picked up during this phase strongly depends on the short wavelength radiative input from the Sun (UV and X-ray). By such processes Venus could have even lost an ocean comparable to that of the Earth's.

Besides the influence of short wavelength radiation on the water loss also solar wind effects have to be taken into account (Bauer, 1983 [22]).

Another important isotope for studying the water-history of Venus is ^{40}Ar. The atmospheric abundance of ^{40}Ar is only ~25% of the radiogenic gas produced inside Venus. This may imply that Venus is not thoroughly degassed and its interior has not been dried over time. A dry interior of Venus can be explained by a near head-on collision of two large planetary embryos. Such a collision would be sufficiently large to melt totally and briefly vaporize a significant proportion of both bodies. This would allow much of the released water to react rapidly with iron (Davies, 2008 [95]).

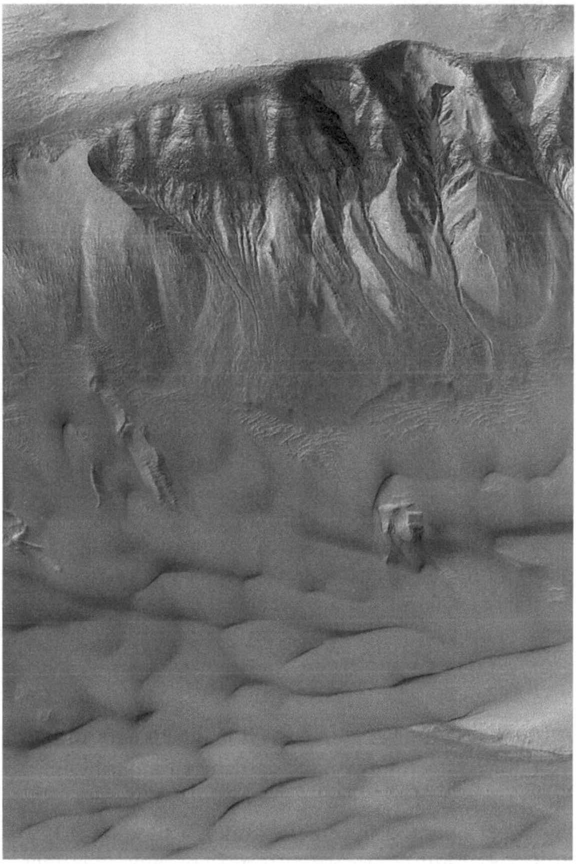

Fig. 3.7 This Mars Global Surveyor (MGS) Mars Orbiter Camera (MOC) image shows a suite of south mid-latitude gullies on a crater wall. Gullies such as these may have formed by runoff of liquid water. The problem is that the temperature is normally too cold and the atmosphere too thin to sustain liquid water on Mars. However, water could burst out from underground layers and remain liquid long enough to erode the gullies. Another explanation of the gullies involve carbon dioxide. Image Credit: NASA/JPL/Malin Space Science Systems

Mars

The story whether there is water on Mars or not and if yes, how much, is not completely finished yet. As we have stated, the debate started more than 100 years ago by the "observations" of the canali which turned out to be artifacts caused by blurring effects in our atmosphere. In 1976 the first unmanned spacecrafts landed on the martian surface (Viking 1 and Viking 2). Mars appeared as a dry, rocky desert with no signs of water.

In the 1990s however, that picture changed again. Many features on the martian surface were detected that could have been only formed by water (as an example see Fig. 3.7).

Estimates of the amount of water outgassed from Mars, based on the composition of the atmosphere, range from 6 to 160 m, as compared with 3 km for the Earth. On the other hand, large flood features, valley networks, and several indicators of ground ice suggest that at least 500 m of water have outgassed. This contradiction can be explained if early in its history, Mars lost part of its atmosphere by impact erosion and hydrodynamic escape (Carr, 1987 [58]).

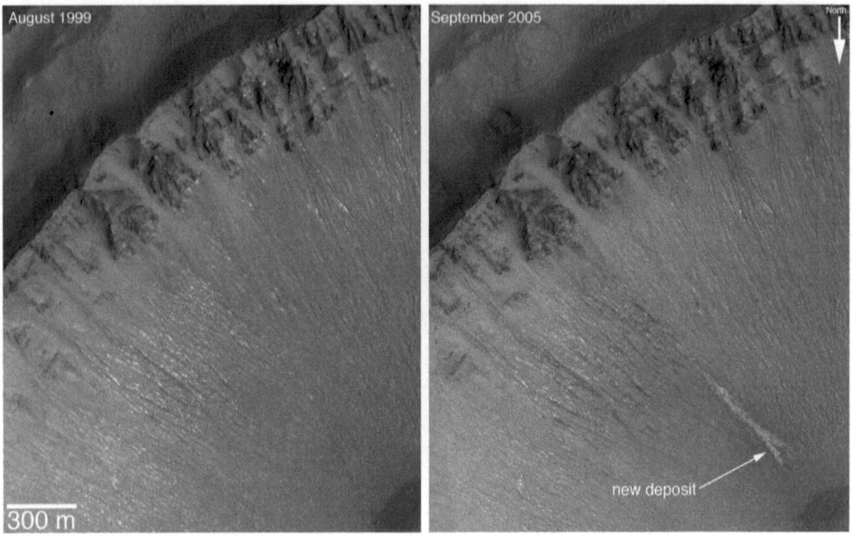

Fig. 3.8 Two images taken by Mars Global Surveyor in Aug. 1999 (*left*) and Sep. 2005 (*right*) of gullies on the wall of a crater in the Centauri Montes region of Mars. The later one shows the appearance of new, light-colored deposit, possibly due to running water. Credit: NASA, Mars Global Surveyor Mission

In 2006, the NASA Mars Global Surveyor orbiter found evidence of water that has flowed fleetingly on the surface. Comparing images of the side of a crater taken in 2001 and 2005, it was seen that the second showed gullies apparently caused by water bursting out of the crater wall (Fig. 3.12. Another example is shown in Fig. 3.8). Gully Formation from Laboratory Simulations were compared with features observed on Mars (Conway et al., 2010 [76]). Flows on the martian surface are erosive, flow faster and further than on Earth, and produce unique sedimentary features (see Fig. 3.9).

Mariner 9 images showed equatorial sinuous channels on Mars. Sagan, Toon and Gierasch, 1973 [288] mentioned two possible stable climates: one that resembles the present day climate on Mars, the other which has atmospheric pressure of about 1 bar. The triggers for a transition from one state to another are changes in the obliquity, solar luminosity and albedo variations of the polar caps.

Climate changes on Mars have been studied by various authors. Evidence for climate change on Mars was investigated by Lewis and Read, 2006 [201]. Abe et al., 2005 [1] used a general circulation model to gain insights on the hydrologic cycle on ancient Mars. They identified four different climate regimes, depending on the obliquity of the spin axis of Mars. (i) warm-upright, (ii) warm-oblique, (iii) frozen-upright, and (iv) frozen-oblique regimes. The period of active hydrologic cycle suggested from the geomorphology on Mars seems to be consistent with that at the warm-oblique regime, which appears at warm (above-freezing) environment with high-obliquity (higher than about 30°) condition. Global circulation models on Mars predict tropical ice accumulations at times of high obliquity (Haberle, 2004 [151]).

Fig. 3.9 Galle crater on Mars with rock layers. Groups of layers cut across each other. Image Credit: NASA/JPL/Malin Space Science Systems

Generally, the obliquity is a primary factor for climate changes. Nakamura and Tajika (2003, [244]) studied obliquity changes on terrestrial like planets that possess a CO_2 atmosphere. Obliquity changes result into drastic climate changes because of a runaway sublimation of permanent CO_2 ice. Their simulation led to ring-like structures of CO_2 ice at mid latitudes.

Yokohata, Odaka, and Kuramoto, 2002 [370] studied the albedo feedback mechanisms of H_2O and CO_2 ices in the atmosphere-ice cap system. High atmospheric pressure presumed for past Mars would be unstabilized if H_2O ice widely prevailed. As a result, a cold climate state might have been achieved by the condensation of atmospheric CO_2 onto ice caps. On the other hand, the low atmospheric pressure, which is buffered by the CO_2 ice cap and likely close to the present pressure, would be unstabilized if the CO_2 ice albedo decreased. Thus there is a strong interaction between the two different ices on Mars, H_2O and CO_2. They contribute to the martian global albedo, a larger value for the albedo means that more incoming solar radiation is reflected back into space and not absorbed in the martian surface, thus the planet is cooled.

There seems to be a lot of ground ice in the martian soil. Figure 3.10 shows deposits on the surface of a crater. This can be deduced from gamma ray measurements. The Mars Odyssey mission arrived at Mars on October 23, 2001 carrying

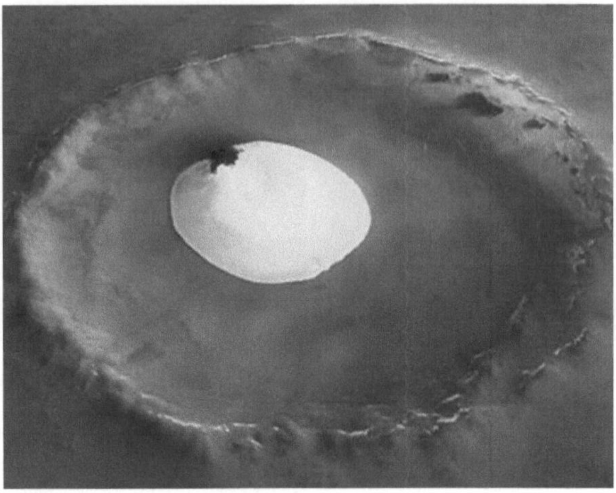

Fig. 3.10 Mars crater with deposits of water ice. Credit: Mars Express mission

such an instrument. A map showing the distribution of water was made based on these measurements. Other observational features such as viscous creep support the hypothesis that water is present in the soil (Fig. 3.11). Mustard et al., 2001 [243] estimate that in the 30 to 60° latitude bands there is ice-occupied soil reservoir that corresponds to a thickness between 1 and 10 m and the reservoir corresponds to $1.5\text{–}6.0 \times 10^4$ km^3. Such a reservoir would be equivalent to a global layer of water 10–40 cm, distributed over the whole martian surface. Their conclusion is that this reservoir has been created during the last phase of large high orbital obliquity less than 100 000 years ago and that it is now diminishing. Let us give a comparison of these numbers with numbers for the Earth. If all ice on Earth melted, the level of the oceans would rise by 64 m.

An example for the search of water on Mars is given in Fig. 3.13. It shows a spectrum obtained with the Mars Exploration Rover Spirit's Moessbauer spectrometer.[12]

Climate changes may also induce oscillations of the obliquity because e.g. when there are large extensive polar caps this influences on the global mass distribution (Rubincam, 1990 [285]).

So Mars may have experienced warm and cold climates episodically in its history. The reasons for climate changes on Mars can be manifold. Besides changes of the obliquity of its spin axis that certainly happened, other effects such as large volcanic eruptions, impacts of large meteorites and others have to be considered (see also Kargel and Strom, 1996 [178].) Climatic change on Mars caused by impact basin formation was studied by Matsui, Tajika, and Abe, 1988 [220].

Martian dust storms and climate change were studied by Leovy, 1986 [199]. Seasonal cycles of dust, carbon dioxide and water in connection with astronomical changes (tilt of obliquity, impacts) have to be considered for modeling the martian climate.

[12] A spectroscopic technique based on the recoil-free, resonant absorption and emission of gamma rays in solids.

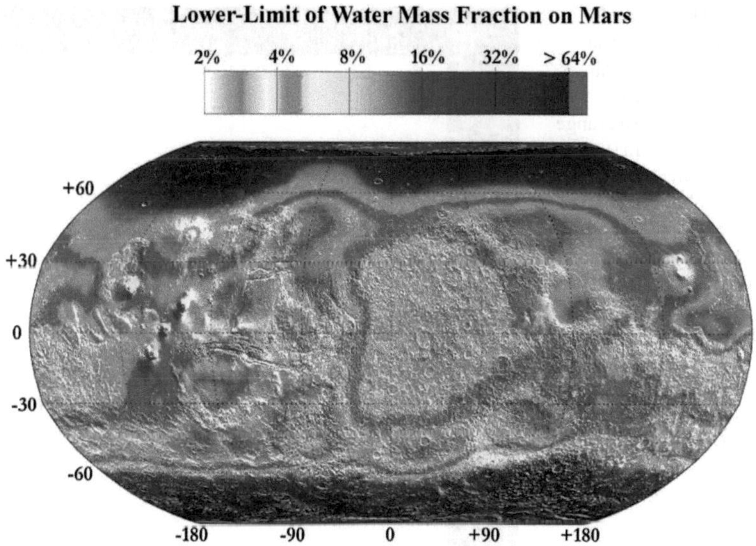

Fig. 3.11 A map showing the distribution of water on Mars. The data were obtained with the Gamma Ray spectrometer, GRS. Credit: Mars Odyssey, GRS Team, LANL, NASA

At present, Mars does not have a global magnetic field. However, in its early phase a dynamo mechanism could have worked causing a global martian magnetic field. Such a field might have effectively shielded the planet's atmosphere from solar wind particles. Later in the history of martian evolution, cooling and/or solidification of the core caused dynamo extinction and exposed the Martian atmosphere to the full solar and galactic particle flux. This might have had considerable influence on the global martian climate (Cordell, 1980 [77]). When the magnetic field of Mars ceased to exist (about 4 Gyr ago), atmospheric escape induced by solar wind began (Dehant et al., 2007 [102]).

Wordsworth et al., 2010 [367] made a three dimensional global circulation model (GCM) of the early martian climate. In their model CO_2 condensation, cloud formation and a water cycle was included. Local water vapor feedbacks compensate reduced CO_2 warming effects. In general CO_2 clouds lead to a substantial warming.

For the water loss in the martian atmosphere also solar wind pick up of ionospheric ions has to be considered (Brecht and Ledvina, 2010 [45]). A detailed study on water loss and evolution of the upper atmosphere and exosphere over martian history was given by Valeille et al., 2010 [344]. The loss of water was studied by taking the loss of atomic oxygen as a proxy. Solar EUV flux enhancements were modeled of 1, 3 and 6 times the present values. Also the solar cycle effect was considered. Maps of ion production by three different processes are given by these authors. The processes involved are: photoionization, charge exchange and electron impact. They estimate that about 10 m of water escaped globally to space over the last ~3.5 Gyr.

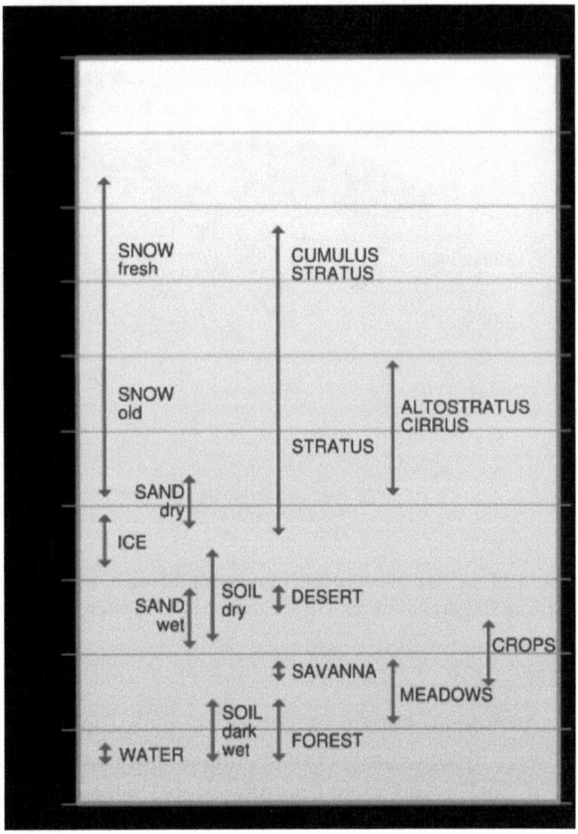

Fig. 3.12 The albedo for different substances. The higher the albedo, the larger the fraction of diffusely reflected sunlight. The range of the albedo is from 0 (*completely dark*) to 1 (*bright*). On Earth this effect is very well known: If a snow covered area warms and the snow melts, the albedo decreases, more sunlight is absorbed, and the temperature tends to increase. If snow forms, a cooling cycle happens. Credit: H. Grohe, Creative Commons BY-SA licence

Geologic History of Mars

The geologic history of Mars can be divided into several epochs:

- Formation: the planet formed within several 10 Myr of the solar system formation and became differentiated into crust mantle and core.
- Pre Noachian period: Mars had probably a magnetic field.
- Noachian period: 4.1–3.8 Gyr BP. Formation of the Hellas Basin.
- End of Noachian: Tharsis formed; hydrous weathering products such as phyllosilicates, extensive sulfate deposits. Warm, wet conditions necessary for fluvial activity were met only occasionally, such as might occur if caused by large impacts or volcanic eruptions.
- Hesperian: volcanic activity continued, 30% of surface was restructured. Large water floods formed episodically, formation of Canyons.
- End of Hesperian: 3 Gyr BP. Geologic activity became weaker.
- Amazonian: Geologic activity confined to Tharsis and Elysium, small floods occurred episodically.

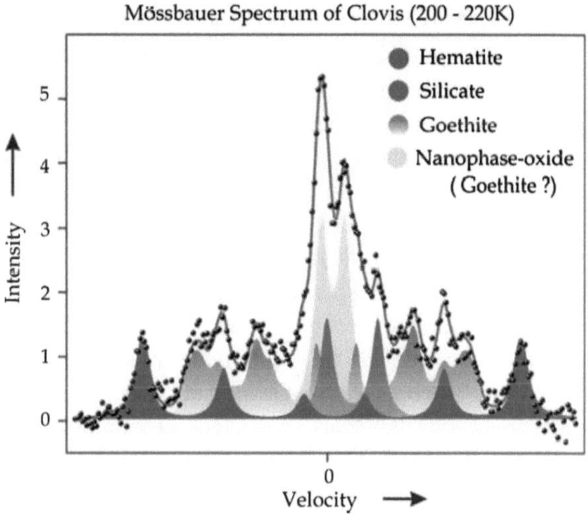

Fig. 3.13 This spectrum, taken by the Mars Exploration Rover Spirit's Moessbauer spectrometer, shows the presence of an iron-bearing mineral called goethite in a rock called "Clovis" in the "Columbia Hills" of Mars. Goethite contains water in the form of hydroxyl as a part of its structure. By identifying this mineral, the examination of Clovis produced strong evidence for past water activity in the area that Spirit is exploring. Image Credit: NASA/JPL/University of Mainz

For more details the reader is referred to Carr and Head, 2010 [59]. There is evidence for recent water flows on Mars based on geologic maps from the Viking mission that indicate geologic and hydrologic activity extending into the Late Amazonian epoch especially in the Tharsis and Elysium area. Examples for such evidence are

- pristine lava flows with few superposed impact craters;
- tectonic features that cut stratigraphically young materials;
- features with possible aqueous origin such as structurally controlled channels;
- spatially varying elemental abundances for such elements as hydrogen (H) and chlorine (Cl) recorded in rock materials up to 0.33 m depth;
- regions of elevated atmospheric methane.

The Solfatara Crater, Italy, has been recommended as a suitable terrestrial analog. Dohm et al., 2008 [106] suggested that the Tharsis/Elysium corridor should be considered a prime target for Mars Reconnaissance Orbiter (MRO)

Fairen et al., 2009 [124] conclude from data from the Mars Exploration Rover (Guseev Crater, impact occurred 2 Gyr BP) that minor amounts of shallow acidic liquid water have been present on the surface of Mars at local scales during the Amazonian Period. They investigated jarosite deposits. Jarosite is a basic hydrous sulfate of potassium and iron with a chemical formula of $KFe_3^{3+}(OH)_6(SO_4)_2$.

A cold wet Mars was described by Fairen, 2010 [123]. A hypothetical martian fluid with a composition resulting from the acid weathering of basalt based on orbiter- and lander-observed surface mineralogy of Mars was included in the models. The simulations show that the hydrological cycle would have been active only in periods of dense atmosphere. A minimum atmospheric pressure is essential for water to flow, and relatively high temperatures (over 245 K) are required to trigger evaporation and snowfall; minor episodes of limited liquid water on the surface could have occurred at lower temperatures (over 225 K). During times with a thin atmosphere and even lesser temperatures (under 225 K), only transient liquid water can potentially exist on most of the martian surface.

Schulze-Makuch et al., 2007 [301] studied possible hydrothermal environment on Mars that could be targets for future Mars missions.

Lobate-debris aprons, lineated valley fill, concentric crater fill, viscous flow features indicate glaciofluvial activity in the Amazonian. These valleys appear qualitatively different from valley networks formed in the Noachian, which can be much longer and often formed integrated networks and large lakes (Fassett et al., 2010 [127]). Andrews-Hanna et al., 2010 [9] investigated Meridiani playa deposits and the sedimentary record of Arabia Terra. They conclude that conditions in the late Noachian to early Hesperian must have been conducive to the existence of liquid water at the surface throughout much of the low latitudes of Mars.

Murchie et al., 2009 [242] made synthesis of Martian aqueous mineralogy from Mars Reconaissance Orbiter data. Aqueous minerals are both diverse and widespread on the Martian surface. Examples are Phyllosilicates, Carbonate-bearing rocks, Hydrated silica occurs with hydrated sulfates. Two to five classes of Noachian-aged deposits containing phyllosilicates and carbonates may have formed in aqueous environments with pH and water activities suitable for life.

3.3 Giant Planets

The four giant planets Jupiter, Saturn, Uranus and Neptune are similar, they have no solid surface and mainly consist of hydrogen and helium with a solid rocky-ice inner core. A review about the giant planets was given by Hubbard, Burrows, and Lunine, 2002 [165].

3.3.1 Jupiter

Jupiter is the largest planet within the solar system (Fig. 3.14). Its mass is 2.5 times the mass of all other planets in the solar system. Jupiter's atmosphere is composed of about 88–92% of hydrogen and 8–12% of helium. There are also traces of carbon, ethane, hydrogen sulfide, neon, oxygen, methane, water vapor, ammonia, silicon-based compound and also phosphine and sulfur. Helium in the Jupiter atmosphere is depleted (compared with the composition of the primordial solar nebula). This can be explained by a precipitation of the helium into the interior of the planet. The composition of Saturn is similar to that of Jupiter, Uranus and Neptune have much less hydrogen and helium.

Jupiter consists of a dense core with a mass of 12 to 45 times the Earth's mass (3–15% the mass of Jupiter). This rocky core is surrounded by dense metallic hydrogen which extends outward to about 78% of the planet's radius. When hydrogen is sufficiently compressed it becomes a liquid of protons and electrons. So this substance possesses a high electrical conductivity which is also typical for metals.

Fig. 3.14 Jupiter; structures in its dynamic atmosphere. At the lower right the great red spot is seen; composed of 4 images taken by NASA's Cassini spacecraft on December 7, 2000. Courtesy: NASA/JPL

Jupiter, Atmosphere

Droplets of helium and neon precipitate through this layer, depleting the abundance of He and Ne in the upper atmosphere. Above the layer of metallic hydrogen the atmosphere consists of liquid and gaseous hydrogen. The temperature strongly varies from the interior to the atmosphere. At the transition, where liquid hydrogen becomes metallic it is 10 000 K and the pressure is 200 GPa.[13] The temperature at the core boundary is more than 35 000 K and the pressure is 3000–4500 GPa.

The clouds of Jupiter are divided in lighter-hued zones and darker belts. The cloud layer is 50 km deep. Underlying the ammonia clouds, there may be a thin layer of water clouds. This could explain lightning detected in Jupiter's atmosphere. Since water is a polar molecule (see Chap. 1), it can carry charge and is capable of charge separation which is needed to produce lightning. Gibbard, Levy, Lunine, 1995 [140]) modeled the thunderstorm separation of electrical charge on precipitating ice particles at varying depths in Jupiter's atmosphere predicting that a condensed-water abundance in the range of at least 1–2 $\mathrm{g\,m^{-3}}$ is required for lightning to occur in jovian thunderstorms. The atmosphere of Jupiter appears in orange and brown colors. The upwelling compounds change color when they are exposed to solar UV radiation. Phosphorus sulfur and hydrocarbons play an important role for the colors.

Summarizing, the main cloud layers in the atmospheres of Jupiter and Saturn are:

[13] 1 Pa = 1 N/m^2 = 100 000 bar = 98 066.5 at. 1 GPa = 10^9 Pa.

Fig. 3.15 Internal structure of Jupiter. In the upper layers the atmosphere changes to a liquid state above a thick layer of metallic hydrogen. In the center there may be a solid core of heavier elements. Credit: NASA, R.J. Hall

- ammonia clouds (150 K),
- ammonium hydrosulfide clouds (200 K),
- water clouds (270 K).

The water clouds are found deepest. Generally the cloud layers in the atmosphere of Saturn are deeper than in Jupiter's atmosphere because of the lower temperature on Saturn.

Observed since 1831, the Great Red Spot is larger than the Earth. It is a persistent anticyclonic storm located south of the equator. The oval rotates counterclockwise with a period of six days. Jupiter has also white ovals (cool clouds) and brown ovals (warmer clouds)—these smaller sized storms can last for hours to months.

Jupiter is surrounded by a faint ring system and the particles of the ring seem to consist mainly of dust not of ice.

As the other gas giants, Jupiter rotates fast and differentially. The zones near the equator rotate faster than those nearer to the poles and Jupiter's shape is that of an oblate spheroid. Jupiter's rotation period is the fastest of all planets in the solar system (less than 10 hours).

The structure of the interior of Jupiter and its atmosphere is illustrated in Fig. 3.15.

3.3.2 Saturn

Saturn is the second largest planet in the solar system and the only planet that is less dense than water ($\rho = 0.69$ g/cm^3) and its mass is 95 Earth masses.

The planet consists of a dense rocky core followed by a layer of liquid metallic hydrogen and above there is a thicker liquid hydrogen and helium layer. The gaseous atmosphere is about 1000 km thick. The core contains 9–22 times the mass of the Earth and its interior reaches a temperature of more than 11 000 K. Saturn radiates 2.5 times more energy into space than it receives from the Sun. This can be explained by a slow compression where gravitational energy is released (Kelvin-Helmholtz mechanism). In addition to that droplets of helium may fall towards its interior releasing heat by friction with the surrounding hydrogen gas.

At the bottom of its atmosphere at a temperature of −23°C there is a layer of water ice clouds (they extend for about 10 km). The ammonium hydrosulfide clouds extend for about 50 km and the temperature drops to −93°C. Ammonia ice clouds extend from about 80 km to regions where the temperature is −153°C. Near the top of Saturn's atmosphere between 200 km to 270 km the atmosphere consists mainly of hydrogen and helium. The strongest known wind systems occur in the atmosphere of Saturn (up to 500 m/s).

Saturn needs about 29,5 years for one revolution about the Sun, its mean distance from the Sun is 9 AU (nine times the distance Sun-Earth). Its rotation axis has a low inclination of 2.4°, so there are no seasonal effects.

The famous rings of Saturn extend from 6630 km to 120 700 km above its surface. The rings are extremely thin, only 20 m and consist mainly of ice: 93% is water ice, 7% is amorphous carbon. The sizes of the ring particles range from micrometers to several m. The gaps observed in the rings are caused by resonances with nearby satellites of Saturn.

The recent Cassini mission[14] has revealed that the ring system possesses an atmosphere like envelope consisting mainly of molecular oxygen, O_2. This molecular oxygen is produced by UV radiation from the Sun that disintegrates the water ice molecules of the ring particles. Loss of water from Saturn's E ring through ion pick-up was investigated by Leisner et al., 2005 [198]. The mechanism studied there is as follows: after being ionized the particles are accelerated by the corotating magnetized plasma. The particles will generate cyclotron waves that were observed by Pioneer 11 and later this was confirmed by Cassini.

As of 2010, 61 moons of Saturn have been identified.

3.3.3 Uranus

On March 13, 1781 Sir William Herschel announced the discovery of Uranus. Together with Neptune this planet belongs to the category of ice giants which is a sub class of the gas giants. Uranus seems to have the coldest atmosphere in the planetary system, the minimum temperature is 49 K (−224°C). Its atmosphere has a layered structure and water ice containing clouds are the deepest clouds. Its distance from the Sun is about 20 times the distance Earth–Sun and it receives only 1/400 of solar radiation than on Earth. The revolution period around the Sun is 84 years. The tilt of its rotation axis is 98°. Each pole gets 42 years of sunlight. Thus the poles

[14]Launched June 2004.

receive greater solar energy input than the equatorial zones however, the zones near the equator are warmer than the polar zones of Uranus.

The mass of Uranus is 14.5 times that of the Earth. Uranus consists of three layers:

- Rocky core at the center; about 0.55 Earth masses, 20% of Uranus' radius.
- Mantle; 13.4 Earth masses; consists of ice, liquid water, ammonia and other substances with a high electrical conductivity. As is the case for Neptune, ice dominates the structure of that planet. Therefore, they are called ice giants.
- Gaseous hydrogen-helium envelope.

Uranus has only a small internal heat source, and it radiates 1.06 times the solar energy absorbed. Maybe, in the early phase of the formation in the solar system, Uranus was hit by a large object which caused the extreme axial tilt we observe today. This impact could have also expelled most of its primordial internal heat.

The atmosphere of Uranus consists of a troposphere followed by a stratosphere and a thermosphere.

The troposphere ranges form −300 km (temperature about 320 K) to 50 km (temperature about 53 K). In the pressure range of 50 to 100 bar water clouds are assumed. In the range of 20 to 40 bar clouds of ammonium hydrosulfide, between 3 and 10 bar hydrogen sulfide and at 1 to 2 bar methane clouds can be found. At the tropopause, the temperature is 53 K, then it increases in the stratosphere due to absorption of solar UV radiation by methane and other hydrocarbons. At the outermost layer of its atmosphere, the temperature is about 800 K.

Uranus has also a ring system (see Fig. 3.16) which consists of very dark particles (the size of the particles ranges from micrometers to several dm). It seems that the rings are young maybe they were formed by impacts on moons of Uranus where matter was shattered.

3.3.4 Neptune

From perturbations of the orbit of Uranus, Le Verrier and Adams made calculations for 1845–1846 estimating the position of this unknown planet and at the observatory of Berlin Galle was the first to observe the planet.

Like Uranus, Neptune is between Jupiter and Earth regarding its mass which is 17 times that of the Earth (this means its mass is 1/19th that of Jupiter). The atmosphere contains about 5–10% of its total mass and extends about 1/10th to 2/10th of its radius.

Methane, ammonia and water are found in increasing concentrations in the deeper atmosphere (Fig. 3.18). There is a mantle (see Fig. 3.17) which is liquid and with temperatures up to 2500 K consisting of a mixture of ammonia, water and methane.[15] This zone therefore is a water-ammonia ocean and at a depth of 7000 km

[15]Though it is liquid, planetary scientists call it icy.

Fig. 3.16 Uranus surrounded by its four major rings and some of its 17 known satellites. This false-color image was generated by Erich Karkoschka using data taken on August 8, 1998, with Hubble's Near Infrared Camera and Multi-Object Spectrometer. Credit: HST, Karkoschka

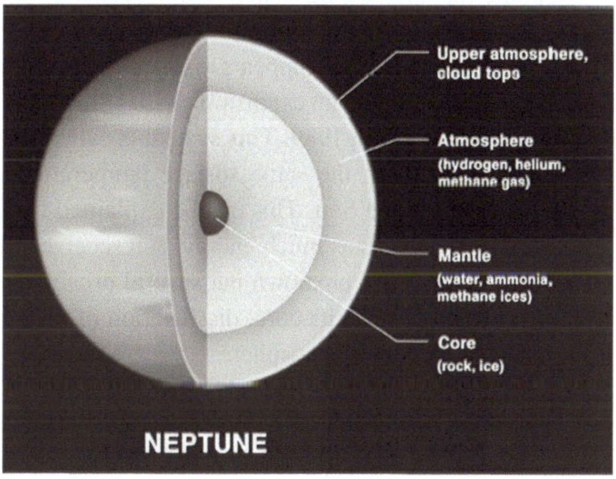

Fig. 3.17 Neptune, internal structure. Courtesy: NASA

Fig. 3.18 Neptune, with some cloud structures (*black spot*). Courtesy: NASA

methane decomposes and maybe diamond crystals precipitate toward the core which consists of nickel and silicates having a mass of about 1.2 times that of the Earth.

Neptune's atmospheric methane predominantly absorbs in the red light, this gives Neptune its bluish appearance. The atmosphere consists of a troposphere, where the temperature decreases and convective motions and clouds can be found and a stratosphere, where the temperature increases. The upper level of clouds are methane clouds (condense at the temperatures found there at a pressure below 1 bar), at pressures about 5 bars ammonia and hydrogen sulfide clouds form and at pressures about 50 bar and temperatures about 0°C there are water clouds.

Neptune has also a ring which mainly consists of silicate or carbon-based ice particles.

The weather on Neptune appears to be more dynamic than that on Uranus. A huge anti-cyclonic storm with an extension of 13 000 × 6000 km was detected by NASA's Voyager 2 spacecraft in 1989. This system is called the "dark spot".

Neptune has about the same surface temperature as Uranus though it is at a greater distance to the Sun. This can be explained by an internal heat source. Neptune radiates 2.6 times as much energy as it receives from the Sun. The sources for its internal heating are unknown but several processes have been proposed such as radiogenic heating from its core, dissociation of methane into hydrocarbon chains, convection in the lower atmosphere.

Because of the 28.3° tilt of its rotation axis, seasonal changes can be observed, but one season lasts about forty Earth years. The planet rotates differential, the equatorial zone rotates with a period of about 18 hours.

3.3.5 Water on Giant Planets

As we have seen water is an important constituent of the giant planets besides hydrogen and helium. This was pointed out by Stevenson and Fishbein, 1981 [326]. The presence of water and other heavy molecules in the interiors of the giant planets may inhibit convection as a means of energy transport there. Guillot, 1995 [150] found that convection is inhibited in Uranus and Neptune when methane reaches an abundance of about 15 times the solar value and in Jupiter and Saturn if the abundance of water is more than about five times the solar value. Roos-Serote et al., 2004 [280] studied the water abundance in the atmosphere of Jupiter by analysing the O/H ratio. This ratio was found compatible with one or more times the solar value. Water vapor is the main source for O (except some small contribution from CO).

Water is present in Jupiter's atmosphere both in the troposphere and in the stratosphere. Observations of tropospheric water were made with the Kuiper Airborne Telescope and the IRIS (Infrared Interferometer Spectrometer and Radiometer) Spectrometer on board Voyager 1 and Voyager 2.[16] From these data the vertical distribution of water in the level from 16 bar in Jupiter's troposphere was derived. A thin H_2O ice cloud would form at 2 bars, $T = 200$ K (Bjoraker, Larson and Kunde, 1986 [30]).

Observation of water vapor from satellite measurements at 557 GHz is reported by Bergin et al., 2000 [23]. The line appears to be formed at maximum pressures of about 5 mbar. Water is not uniformly mixed but increases with altitude above the condensation level. In Jupiter and Saturn, the amount of water implied by the data is 1.5–2.5 times larger than inferred from Infrared Space Observatory data.

IRIS also revealed stratospheric water ice clouds (Simon-Miller, 2000 [311]).

The meteorology of Jupiter is quite complex because there exist two energy sources: sunlight and a poorly understood internal heat source. Even giant storms are observed in Jupiter's atmosphere. Gierasch et al., 2000 [141] observed a storm with lightnings with a vertical extend of 50 km and a length of 4000 km. They conclude that moist convection-similar to large clusters of thunderstorm cells on the Earth-is a dominant factor in converting heat flow into kinetic energy in the jovian atmosphere, thus water vapor plays an important role in Jupiter's atmosphere.

Abundance measurements of oxygen in Jupiter's atmosphere suggest that during its formation phase carbonaceous matter played a more important role than condensed ice. Therefore, Jupiter (and possibly Saturn) are found at the carbonaceous condensation/evaporation front (the "tar line") and the snow line was located farther out in the solar nebula (Lodders, 2004 [206]). This could also explain why Galileo probe measurements found much less water in Jupiter's atmosphere than expected. The snow line denotes the minimum radius from the Sun at which water ice could have condensed, at about 150 K, so it is the point in the solar nebula where water ice condenses. The tar line denotes the point where asphalt or tar-like material formed, pushing the snow line farther out in the solar nebula.

---

[16] The voyager mission consisted of two satellites launched in 1977 that reached Jupiter in 1979, Saturn in 1981 and Voyager 2 reached Uranus in 1986 and Neptune in 1989.

Chemical constraints on the water abundance in the deep atmosphere of Saturn were given by Visscher and Channon, 2005 [346]. Saturn is known to develop the largest scale convective storms in the Solar System. Water storms that also occur in the atmosphere of Jupiter may develop velocities up to 150 m/s (Hueso and Sánchez-Lavega, 2004 [167]).

Atreya, Baines, and Egeler, 2006 [12] investigated the existence and observational hints for an ocean of water-ammonia on Neptune and Uranus. Observational hints are from their tropospheric cloud structure and from the existence of a magnetic field of these planets that must be maintained by a dynamo action that could take place in such a water-ammonia ionic ocean.

The influx of water from the Saturnian rings and or its satellite Enceladus may explain the observed radio occultation profiles of electron densities in Saturn's ionosphere (observations made by Pioneer 11 and Voyager) (Mueller-Wodarg, Mendillo, and Moore, 2006 [239]). The influx of water from the rings of Saturn was already discussed by Connerney and Waite, 1984 [75]. In this paper the following numbers for the water influx are given:

- A planet-wide influx of about 4×10^7 molecules $cm^{-2}\,s^{-1}$ of water from the rings is consistent with the observed ionospheric electron densities.
- An enhanced influx of water occurs at latitudes (−38, +44 deg) connected magnetically at the inner edge of Saturn's B ring, where an electromagnetic erosion process takes place.
- The present-day influx at these latitudes may be as large as 2×10^9 molecules $cm^{-2}\,s^{-1}$.

3.4 Dwarf Planets

In August 2006 the International Astronomical Union defined a new class of objects in the solar system, the dwarf planets. These objects must fulfill the four conditions:

1. Orbit around the Sun.
2. Sufficient mass, that implies a hydrostatic equilibrium and an almost spherical shape.
3. Has not cleared the neighborhood around its orbit.
4. Is not a satellite.

Concerning their diameters, Ceres (974 km), Pluto (2306 km), Eris (1150 km), Makemake (1500 km) and Haumea (2400 km) are considered as real dwarf planets. Objects with icy cores need a smaller diameter (about 400 km) to relax into gravitational equilibrium. More than 40 objects fall into this subclass also called Plutoids.

Pluto was discovered in 1930 and first designated as the 9th planet of the solar system. Since the conference of the International Astronomical Union in Prague, 2006, it is classified however as a dwarf planet together with other recently discovered similar objects. Also Ceres, which was discovered at the beginning of 1800, and classified as the first asteroid between the orbits of Mars and Jupiter, now belongs to this type of objects. Over the last two decades several similar objects like Pluto or Ceres were found.

For example, the diameter of Pluto and Eris is nearly the same, about 2400 km, the semi major axis of Pluto is 39.5 AU (high eccentric, orbital radius between 29.66 and 49.3 AU), that of Eris 67.67 AU (high eccentric, orbital radius between 37.77 and 97.56 AU). Ceres has a diameter of 975 × 909 km and the semi major axis is 2.76 AU. Pluto is also regarded as a prototype of *Trans Neptunian Objects* (*TNOs*).

3.4.1 Pluto

Due to the eccentricity of its orbit, Pluto is closer to the Sun than Neptune for about 20 years. Pluto crossed Neptune's orbit from 1979 to 1999. As Pluto approaches its perihelion,[17] it reaches the maximum distance from the ecliptic $\sim 17°$. Therefore, a collision with Neptune cannot happen and the closest possible distance between the two planets is about 18 AU.

Pluto's rotation period is 6.387 days. Similar to Uranus, Pluto rotates with its poles almost in its orbital plane. Pluto's rotational axis is tipped 122 degrees. When Pluto was first discovered, its relatively bright south polar region was the view seen from the Earth (see also Fig. 3.19). Pluto appeared to grow dim as our viewpoint gradually shifted from nearly pole-on in 1954 to nearly equator-on in 1973. With the Hubble Space telescope, it was possible to detect up to 12 distinct regions on Pluto's surface. In Fig. 3.19 opposite hemispheres are shown. The different bright and dark features might be impact basins. The prominent northern polar cap is produced by different frost deposits.

Synoptic observations of 134340 Pluto from the Hubble Space Telescope were reported by Storrs, and Eney, 2010 [327]. They reconstructed images over almost two decades to look for surface variations.

Pluto has a large satellite, Charon, which was discovered in 1978 (Fig. 3.20). A comparison between these two objects shows that Charon might be the result of a large impact on Pluto (like the formation of the Moon). Pluto and Charon are gravitationally locked. They always keep the same hemisphere to each other. The average distance between Pluto and Charon is only 19 570 km. The average density of Charon is 1.65 $\mathrm{g\,cm^{-3}}$ and its mass is 0.11 that of Pluto.

Two additional satellites of the Pluto-Charon system were detected in 2005 on images taken with the Hubble Space telescope: Nix and Hydra. They are small and not spherical. The diameter of Nix is between 44 and 130 km and the semi-major axis of its orbit is 48 600 km. Hydra is slightly more distant to Pluto, the semi-major axis of its orbit is 65 000 km and its diameter is estimated between 110 and 160 km.

Between 1985 and 1990 mutual eclipses were observed because during that period the Pluto-Charon orbit was seen edge on from the Earth. These eclipses enabled

[17] This is the point closest to the Sun.

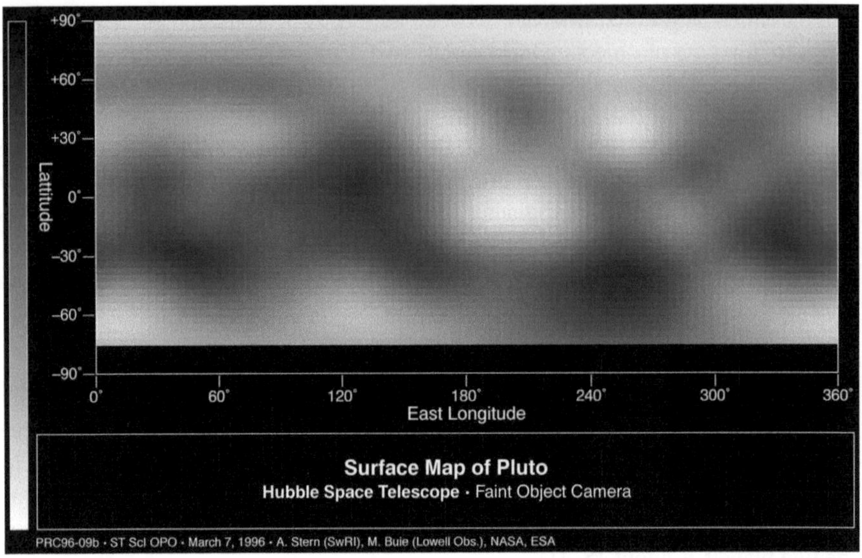

Fig. 3.19 The surface of Pluto is resolved in these NASA Hubble Space Telescope pictures, taken with the European Space Agency's (ESA) Faint Object Camera (FOC) aboard Hubble Space Telescope. Courtesy: NASA/ESO

Fig. 3.20 Pluto with its large satellite Charon. Courtesy: NASA/ESO

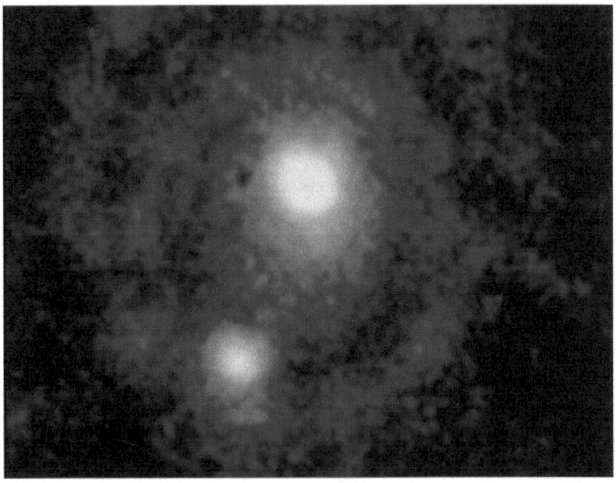

to investigate the combined spectra of Pluto and Charon and later to subtract the individual spectra from the combined spectrum and therefore to determine the surface composition. Charon's diameter is 1207 km, half that of Pluto (Marcialis, Rieke and Lebofsky, 1987 [216]).

Owen et al., 1993 [256] studied surface ices and composition of Pluto by observations in the 1.4 to 2.4 micrometer range. They claimed that frozen nitrogen is more abundant than other ices. Analyzing spectra of the two objects from Pluto and Charon in the range from 1.0 to 2.5 micrometers showed clearly that:

- The spectra are different.
- The spectrum of Charon shows the signature of crystalline water ice and also ammonia ices.
- Mixtures of ammonia ices with water ices could be the origin of surface features (flowlike) that were observed on icy satellites.

Evidence for crystalline water and ammonia ices on Pluto's Satellite Charon was given by Brown, Calvin and Wendy, 2000 [46] and Brown and Calvin, 2000 [46].

In 2007 the Gemini observatory detected patches of ammonia hydrates and water crystals on Charon's surface . This may be explained by active cryo-geysers.

Protopapa, Herbst and Böhnhardt, 2007 [268] observed reflectance spectra of Pluto and Neptune's satellite Triton. Besides the known absorption bands of methane ice, they found absorption bands centered at 4.0 and 4.6 microns. This can be explained as a signature of CO. The formation of substances like HCN and HNC and their spectral signatures are studied by Moore and Hudson, 2003 [235] and also substances like HCN, HNC, HNCO, NH_3, NH_4OCN, and NH_4CN that are involved in the syntheses of biomolecules such as amino acids and polypeptides.

Since Charon is largely an icy body, it is assumed to consist of less rock than Pluto formed from material of the icy crust of Pluto. It is interesting that ice seems to be in crystalline form. Under normal conditions, solar radiation would have degraded ice to an amorphous state within about 30 000 years. So this ice must have been recently deposited.

3.4.2 *Ices on Other Dwarf Planets*

The dwarf planet 90377 Sedna was detected in 2004 and its diameter is estimated about 1800 km. Its orbit is highly eccentric: semi-major axis 484 AU, the perihelion distance is 76 AU, the aphelion distance 892 AU. It has a red color, similar to that of Mars. This can be explained either by a concentration of iron oxides or organic compounds on its surface.

Sedna is, so far, the largest and most distant trans-neptunian object. It was observed with the two 8.2 m VLT telescopes of ESO (Cerro Paranal, Chile). From its spectrum it was deduced that more than 50% of the surface could be covered with ice and the presence of N_2 and CH_4 was confirmed. One revolution of Sedna takes more than 10 500 years but during about 200 years, when this object is close to the Sun, a thin atmosphere of N_2 may form (Barucci et al. 2005, [20]).

Chapter 4
Satellites of Planets in the Solar System

The satellites of the giant planets are of particular interest for the search of water and therefore also for astrobiology in general. A review about the evolution of the icy satellites of the giant planets has been given by Schubert et al., 2010 [298].

4.1 Galilean Satellites

The four largest satellites of Jupiter, Io, Europa, Ganymede and Callisto can be seen using a binocular. In 1609, Galilei was the first to observe them through a telescope. Because of their orbital motions, they appear in the equatorial plane east and west of Jupiter. Some important physical parameters of the Galilean Satellites are given in Table 4.1.

In Europe, there is a study (HADES) for a satellite mission to Europa (orbiter, lander, cryobot) (Böttcher et al., 2009 [39]).

4.1.1 Io

It was a big surprise when on *Io* active volcanism was detected[1] on its surface (see Fig. 4.1). On Earth, the heat source that produces volcanic activity comes from its interior where radioactive materials decay and release energy and also from heat left over from its formation, accretion heat. This however cannot explain the volcanism on Io. The satellite is too small and should have cooled out. The only mechanism to maintain active volcanism on Io is tidal heating (see also Moore, 2003 [236]). The semi major axis of the orbit of Io is 421 800 km so it is close to Jupiter and exposed to strong tidal forces. One revolution around Jupiter takes only 1 d 18 h 27 min. The albedo of Io is 0.61 which is a relatively high value. Other parameters

[1] 1979, Voyager 1.

A. Hanslmeier, *Water in the Universe,* Astrophysics and Space Science Library 368,
DOI 10.1007/978-90-481-9984-6_4,

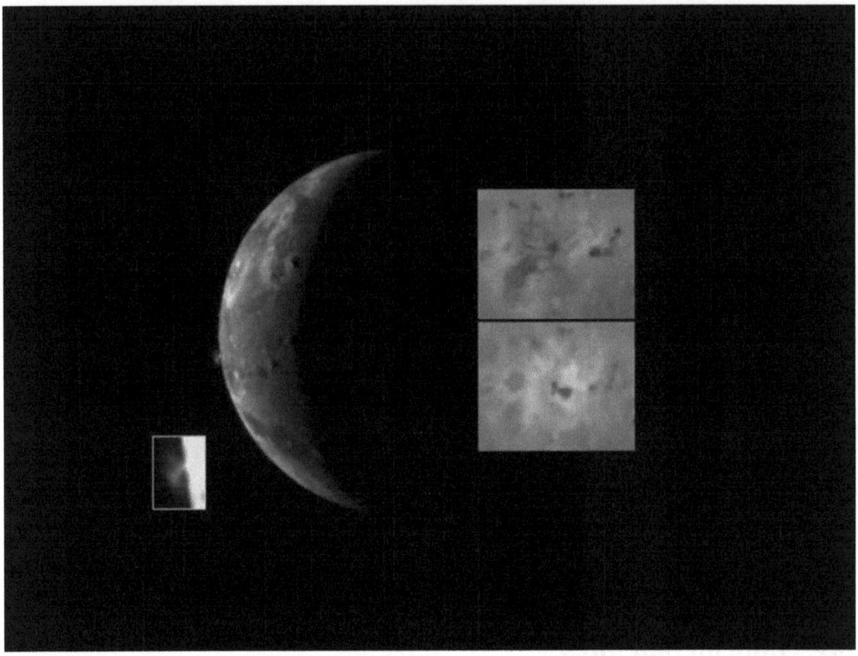

Fig. 4.1 The Jupiter satellite Io: due to tidal heating the body with the most active volcanism in the Solar System. Courtesy: NASA/Galileo mission, JPL

Table 4.1 The four Galilean satellites

Satellite	Diameter km	Mass kg	Density g/cm^3	Semi-major axis km
Io (J I)	3660	8×10^{22}	3.53	421 800
Europa (J II)	3121	4.8×10^{22}	3.01	671 000
Ganymede (J III)	5262	1.5×10^{23}	1.94	1 070 400
Callisto (J IV)	4820	1.08×10^{23}	1.83	1 882 700

are given in Table 4.1. The sidereal rotation period is exactly the same as its period of revolution as a result of the tidal interaction.

On the surface of Io 400 active volcanoes and more than 150 mountains have been detected. There are volcanic plumes that extend up to 500 km. The tidal forces of Jupiter on Io are 6000 times stronger than those of the Moon on Earth. In addition to these forces, there are also tidal forces caused by the two other Galilean satellites *Europa* and *Ganymede* (these are comparable to the tidal force of the Moon). Furthermore, the strength of these forces varies because the orbit of Io is elliptical. The variation of the tidal forces of Jupiter due to Io's elliptical orbit is 1000 times the strength of the tidal forces of the Moon.

On Earth, the tidal force cause deformations of the whole Earth's crust between 20 and 30 cm. On Io, the deformations can reach up to 300 m.

Water on Io

As we will show later, there are several spectral signatures for the presence of ices consisting of H_2O and SO_2. Sulfur dioxide, SO_2 has a melting point of $-72.4°$ and a boiling point of $-10°C$. It is a bent polar molecule. Due to its antimicrobial properties it is used as preservative for fruits and serves also as an antibiotic and antioxidant, protecting wine from spoilage by bacteria and oxidation.[2]

Water is escaping from Io which can be explained by different mechanisms:

- Thermal escape
- Photolysis
- Gas-phase charged particle interactions
- Charged-particle sputtering.

Pilcher, 1979 [262] demonstrated that charged particle sputtering maybe an effective mechanism for the removal of water ice from Io's surface. Another important mechanism could be vaporization of ice by meteoroid impacts. Some small amount of water maybe still present on Io's surface. Spectra in the IR in the 2.5–5.0 micron region exhibit bands that can be accounted for by the presence of SO_2 ices that contain about 3% H_2S and 0.1% H_2O (see Salama et al., 1990 [291]). They form by a condensation on a surface at a temperature of 100 K. Using data from the Kuiper Airborne Observatory,[3] a strong absorption band at 2.79 microns was discovered which is also indicative for ices consisting of SO_2 and H_2O (Salama et al., 1994 [292]). Laboratory reflectance spectra of SO_2 frost and ice and thin H_2O frost show that the observed 2.79 micron band can also be explained by the presence of SO_2 frost alone (Nash, 1994 [245]), so there is some controversy about the interpretation of the spectral signatures. The ratio of SO_2 to H_2O ice was investigated by Dahmani and Khana 1996, [90] and their conclusion was that H_2O must be far less than 1 part in 10^5 SO_2.

The details of the spectra described above also tell us that the SO_2 is about 30% more abundant on Io's leading side than on its trailing side and that the ices are relatively well crystallized, i.e. in form of ice cubes.

4.1.2 Europa

Europa has a mean orbital radius of 671 079 km and its orbital period is 3.55 days. The radius is 1560.8 km and the rotation period is synchronous to the orbital pe-

---

[2]Also known by the designation E220.

[3]This was an observation platform mounted on highly modified C-141A jet transport aircraft to observe in the infrared at a height of up to 14 km. With that instrument the atmosphere of Pluto was found in 1988.

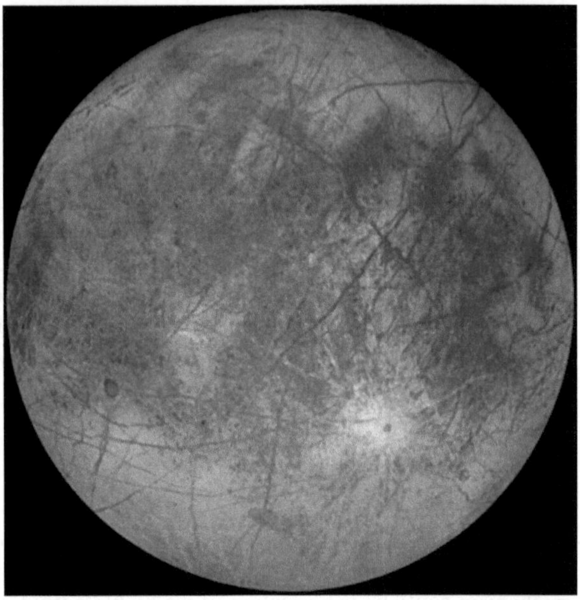

Fig. 4.2 The Jupiter satellite Europa: due to tidal heating beneath an ice curst an ocean of liquid water is highly probable. Courtesy: NASA/Galileo mission, JPL

riod. The surface is extremely smooth and the albedo is 0.64 because the surface is composed of ice (Fig. 4.2). On the surface one observes dark streaks over the entire globe. It is supposed that they have been produced by a series of volcanic water eruptions or geysers. Since Europa is tidally locked to Jupiter, a certain pattern of these structures should have evolved (Fig. 4.3). It seems however that this is only valid for the youngest patterns. The older the patterns, the greater the difference from this arrangement. This can be explained if Europa's surface rotates faster than its interior. A subsurface ocean could decouple the motion of the ice crust from that of the rocky mantle. The temperature at the surface of Europa is 110 K. But due to tidal interaction under the surface there could be a liquid layer. The outer crust of solid ice is believed between 10 and 30 km thick. The underlying ocean could be about 100 km thick.

The possibility that tidal dissipation in an ice crust on Europa preserved a liquid layer of water beneath it was discussed Cassen, Reynolds, and Peale, 1979 [60]. Further evidence for liquid water below the ice layer comes from a restructuring of Europa's surface. Some observational evidence could be insulating layers of frosts derived from water boiling up between cracks in the surface crust (see Squyres, Reynolds, and Cassen, 1983 [322]). Crawford and Stevenson, 1985 [83] discussed gas-driven water volcanism and resurfacing on Europa. Such a phenomenon can be explained by assuming a relatively thin ice shell that is subjected to elastic fracture. The subsequent eruption is first dominated by gas and then a less extended foam eruption (Crawford and Stevenson, 1988 [84]). Greenberg et al., 1999 [145] investigated the formation of chaotic terrain by melt-through from below, and tectonic terrain (cracks, ridges, and bands) by tides have probably been the two dominant

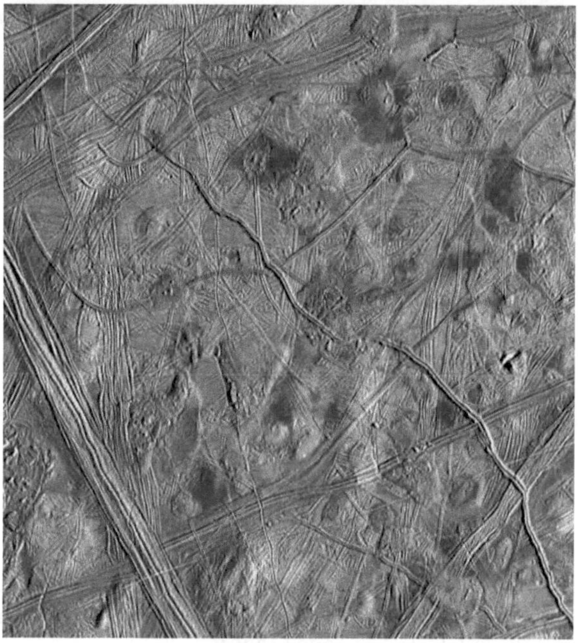

Fig. 4.3 Surface features on Europa. Courtesy: NASA/Galileo mission, JPL

diachronous[4] resurfacing processes over Europa's geological history. Gaidos and Nimmo, 2000 [137] also found that some surface features of Europa are formed by soft ice that is heated by viscous dissipation of tidal motion along faults.

The thickness of the ice shell is still not clear. One method to estimate this value is to observe rotation variations (librations) of Europa. Calculations show, that the presence of an ocean increases the amplitude of libration by about 10%, depending mainly on the thickness of the icy shell are to be expected (van Hoolst et al., 2008 [345]).

Is it possible to observe a subsurface ocean of Europafrom measurements made at its surface? Khurana et al., 2002 [184] showed how electromagnetic and seismic observations made from as few as two surface observatories comprising a magnetometer and a seismometer offer the best hope of unambiguous characterization of the three-dimensional structure of the ocean and the deeper interior of this moon.

Sotin, Head and Tobie (2002, [313]) showed that the ice viscosity on Europa in connection with tidal heating will lead to heating of rising diapirs.[5] This would result in a disruption of near-surface material and the formation of lenticulae and chaos. Also a relatively thick ice sheet of 20 km overlying an ocean might produce such effects.

[4]This term denotes a geologic formation in which apparently similar material varies in age from place to place.

[5]In geology, diapirs commonly intrude vertically upward along fractures or zones of structural weakness through denser overlying rocks.

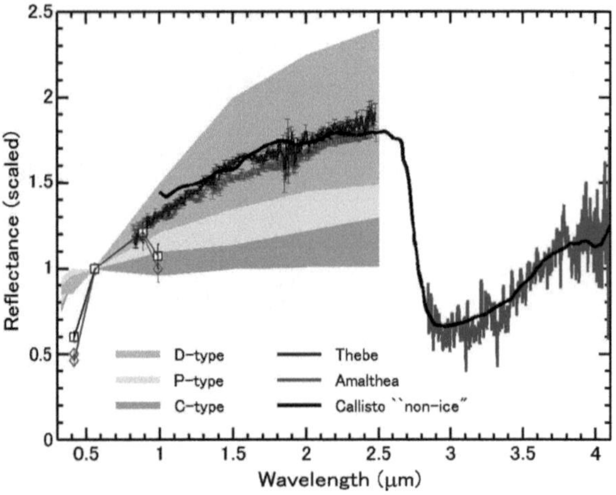

Fig. 4.4 A comparison of the reflectivity of Amalthea, Thebe, Callisto and Asteroids. Amalthea and Thebe have reflective spectra similar to those seen in regions of Callisto where there is little water ice (*black line*). The dip in the spectrum around 3 μm indicates the presence of water containing minerals. Copyright Subaru Telescope, National Astronomical Observatory of Japan (NAOJ)

Evidence for a warm subsurface ocean comes from the forms of chaotic terrain and the estimate of the age of the surface (Melosh et al., 2002 [227]). In this context the term "warm" means that most of the ocean is at the temperature of maximum density (Melosh et al., 2004 [228]).

With the likely presence of water oceans relatively close to its surface, coupled to possible sources of organics, the emergence and sustaining of life on Europa seems possible and this was pointed out by Raulin, 2005 [273]. Goguen et al., 2006 [143] studied the rate of decomposition of four amino acids by UV-photolysis in a mm thick crystalline water ice at $T = 100$ K. From their lab experiments they conclude that half of these amino acids in the upper meter of low latitude ice on Europa will be decomposed by solar UV on a 10 year timescale. Photons between 160 and 300 nm wavelength are responsible for this decomposition.

In order to explain the surface structures on Europa, cryovolcanism plays an important role. A cryovolcano is an icy volcano erupting erupt volatiles such as water, ammonia or methane. Cryovolcanoes form on low temperature objects in the solar system such as on the icy moons.

Eruptions on Europa's surface could be similar to basaltic eruption on Earth: $\lesssim 100\ \mathrm{m^3\,s^{-1}\,m^{-1}}$ flow lengths of tens of hundreds of km and the turbulent nature of the flow motion near the vent several tens of meters of thermal erosion in the proximal parts of water flows (Wilson and Head, 1998 [361]). The existence of hydrothermal vents, similar to those on the ocean floor's on earth is highly probable and this may have led to formation of primitive forms of live.

In Fig. 4.4 a comparison of the reflectivity of Amalthea, Thebe, Callisto and Asteroids is given (obtained with the 8.2 m Subaru Telescope). Thebe is a small satellite of Jupiter (55 × 45 km), Amalthea is one of the inner satellites of Jupiter (mean diameter about 165 km). In this figure the signature of water is clearly seen. Amalthea radiates more energy than receiving, maybe it is heated by electric currents in Jupiter's magnetosphere.

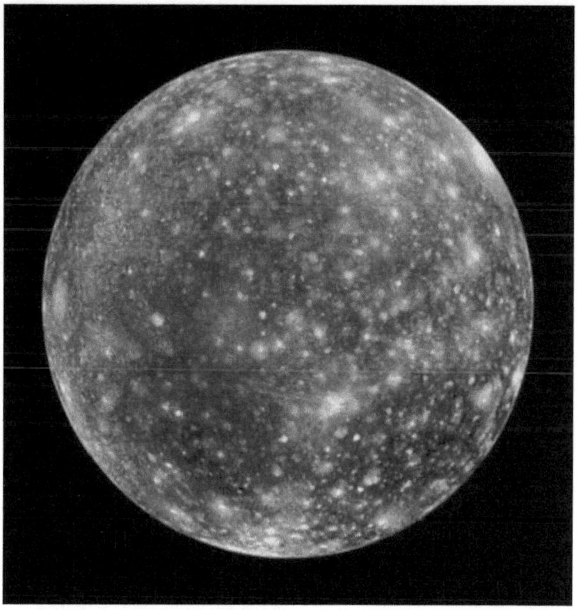

Fig. 4.5 Callisto. Courtesy: NASA/Galileo mission, JPL

4.1.3 Callisto

Jupiter's second largest satellite may also harbor a liquid salty ocean under its icy cratered crust. This is based on the detection of changes of the magnetic field of Callisto (changes in Europa's magnetic field also gave a first hint to a liquid ocean). These changes can be explained by assuming varying electrical currents associated with Jupiter that flow near Callisto's surface (Fig. 4.5). The only source where such currents can exist in Callisto is a layer of melted ice underneath. If such a liquid layer were salty like Earth's oceans then electrical currents could be efficient enough to produce the field and the variations in it. One possibility would be that the water contains up to 5% of ammonia (see Spohn and Schubert, 2003 [318]). But there is one difference between Callisto and Europa. The mean radius of the orbit is 1.88×10^6 km. Thus the tidal forces due to Jupiter are much smaller. Kuskov and Kronod, 2006 [191] claim that Callisto must be partially differentiated into an outer ice, ice I-crust, a water ocean, a rock-ice mantle and a rock-iron core mixture. They estimated the thickness of the ice-I-crust between 135–150 km and that of the underlying water layer between 120 and 180 km.

4.1.4 Ganymede

Also the possibility of a subsurface ocean on Ganymede, the largest satellite in the solar system is discussed. Ganymede participates in a 1:2:4 orbital resonance with

Fig. 4.6 The Jupiter satellite Ganymede. True-color image taken by the Galileo probe. Courtesy: NASA/Galileo mission, JPL

the satellites Europa and Io, respectively. It is larger in diameter than the planet Mercury but has only about half its mass which is indicative for its relatively low mean density. This satellite is mainly composed of silicate rock and water ice. Ganymede is differentiated with an iron core. At a depth of about 200 km a saltwater ocean is assumed.

A third of its surface (Fig. 4.6) is covered with dark terrains which are saturated with impact craters. The lighter regions show also other features like grooves and ridges and are much younger. Tectonic activity might have formed them.

Ganymede's magnetic field that is generated by a dynamo process is similar to our Earth's field.

The surface albedo is about 0.43. This indicates the presence of water ice on its surface. Near-infrared spectroscopy has revealed the presence of strong water ice absorption bands at wavelengths of 1.04, 1.25, 1.5, 2.0 and 3.0 μm. Like the surface of our Moon, Ganymede's surface is asymmetric. The leading hemisphere is brighter than the trailing one.

Ice in the interiors of Ganymede and Callisto was discussed by Schubert, Ellsworth and Stevenson, 1980, [297]. Certain surface features on Ganymede (gently sloping dome-shaped features) maybe explained by water volcanism driven by a major impact, Squyres, 1980 [321]. Allison and Clifford, 1984 [6] made a study on ice-covered volcanic water flows on Ganymede. As it was pointed out already, at 3 micron there is a water ice absorption feature in the spectra. Measurements of this absorption feature demonstrated that a 50% real ice coverage can be expected for the ice on Ganymede and only 10% for Callisto (Spencer, 1987 [315]). The grooved terrain of Ganymede's surface can be explained by eruption of liquid H_2O

magmas (Allison and Clifford, 1987 [7]). In this paper the thermal evolution of such a flow is considered in detail. They give the following numbers:

- The minimum unfrozen lifetime of a 5-m flow is approximately 12.5 days while a 10-m flow would survive for at least 50 days.
- Heating resulting from frictional heat loss could reasonably extend these lifetimes by 50 percent or more.
- With a discharge rate of the order of 1–10 km^3/day, an individual volcanic water flow could flood an area of about 10 000 km^2 before freezing.

Large parts of Ganymede are resurfaced and volcanic eruptions of liquid water or solid ice to tectonic deformation may explain these processes. Data from the Voyager and Galileo spacecraft[6] reveal bright, smooth terrains that lie at roughly constant elevations 100 to 1000 m below the surrounding rougher terrains (Shenk et al., 2001 [307]). The CO_2/H_2O ratio was investigated by Hibbitts, Hansen and McCord, 2002 [162]. Reflectance spectra returned by the Near Infrared Mapping Spectrometer (NIMS) aboard the Galileo spacecraft show that the water-bearing non-ice material on Ganymede hosts some of the CO_2, but on Callisto an anhydrous OH-bearing material hosts the CO_2.

4.2 Satellites of Saturn

4.2.1 Overview

The main physical parameters of Saturn'ssatellites are given in Table 4.2.

Considering the relative masses of Saturn's moons we have to state that 96% of the mass of them is concentrated in Titan, thus this satellite dominates the system. The six medium-sized spheroidal moons constitute 4%, and all small moons together just 0.01%. As of 2006 sixty satellites of Saturn have been confirmed.

Table 4.2 Some important parameters of the largest moons of Saturn

Moon	Diameter km	Mass kg	Orbital radius km	Orbital period d
Mimas	400	0.4×10^{20}	185 000	0.9
Enceladus	500	1.1×10^{20}	238 000	1.4
Tethys	1060	6.2×10^{20}	295 000	1.9
Dione	1120	11×10^{20}	377 000	2.7
Rhea	1530	23×10^{20}	527 000	4.5
Titan	5150	1350×10^{20}	1 222 000	16

[6]The Galileo spacecraft was launched in October 1989; it arrived at Jupiter in 1995; on September 21, 2003, the mission was terminated by sending the orbiter into Jupiter's atmosphere at a speed of nearly 50 km/s to avoid any chance of it contaminating local moons with bacteria from Earth.

Fig. 4.7 A comparison between the Earth, the Moon and Titan. Photos are taken from NASA

4.2.2 Titan

Titan is Saturn's largest satellite and it is the second largest satellite in the solar system. This satellite was detected in 1655 by Ch. Huygens. Its mass is 1.35×10^{23} kg. The mean distance from Saturn is 1.22×10^6 km and Titan's orbital period equals its rotation period, $P \sim 15.94$ d. Titan's equatorial radius is 2575 km. The Galilean satellite Ganymede has a diameter of 5260 km and is slightly larger. For illustrative purposes, a comparison between Titan, the Moon and the Earth is shown in Fig. 4.7.

Due to the large solar distance of the Saturnian system, Titan's mean surface temperature is about -178°C. One of the most interesting facts about Titan is that its surface is completely hidden by a very dense atmosphere, the mean atmospheric pressure is about 1.5 bar.

Titan is the only satellite known in the solar system to have a such dense atmosphere. By diameter and volume it is larger than planet Mercury, but only half as massive. This implies that this planet like satellite is mainly composed of rock and ice. There have been many speculations about the nature of its surface. It was suspected that there might be surface structures similar to that found on Earth. Its atmosphere was detected by J. Campos Sola from measuring the limb darkening in 1903. 98.4% of its atmosphere is nitrogen, the remaining 1.6% are composed mainly of methane, CH_4 and other trace gases. Therefore, Titan and Earth are the only two objects in the solar system that contain a nitrogen rich atmosphere. Among the trace gases we find ethane, diacetylene, methylacetylene, acetylene, propane, cyanoacetylene, hydrogen cyanide, carbon dioxide, carbon monoxide, cyanogen, argon and helium.

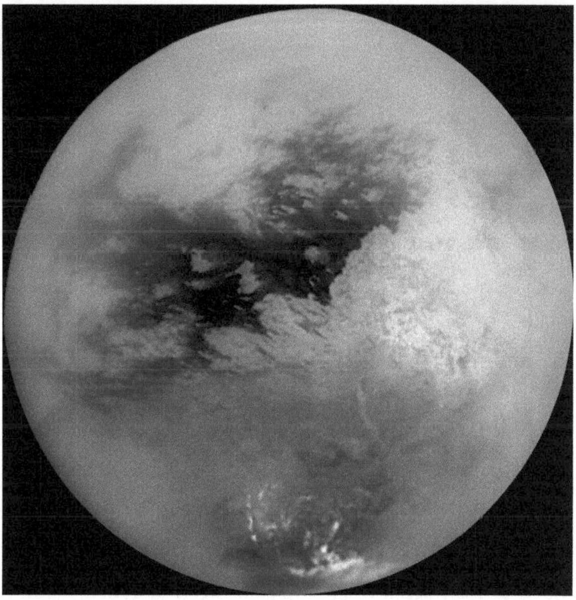

Fig. 4.8 A mosaic of nine processed images acquired during Cassini's first very close flyby of Saturn's moon Titan on Oct. 26, 2004. The brighter region on the right side and equatorial region is named Xanadu Regio. NASA/JPL/Space Science Institute

Due to the solar UV irradiation, methane breaks up in its higher atmosphere and various hydrocarbons are formed.

Four NASA spacecraft have been sent to explore Saturn. Pioneer 11 was first to fly past Saturn in 1979. Voyager 1 flew past a year later, followed by its twin, Voyager 2, in 1981. The Cassini spacecraft is the first to explore the Saturn system of rings and moons from orbit and it was launched on October 15, 1997. It is interesting to mention the complicated cruise to Saturn: Cassini performed two gravity-assist flybys of Venus on April 26, 1998 and June 24, 1999. On August 18, 1999 Cassini did a gravity-assist flyby of Earth. An hour and 20 minutes before closest approach, Cassini made the closest approach to the Moon at 377 000 km, and took a series of calibration images. On Jan. 23, 2000, Cassini performed a flyby of Asteroid 2685 Masursky. Cassini entered orbit about Saturn on Jun. 30, 2004. The Cassini orbiter was built and managed by NASA's Jet Propulsion Laboratory. In Oct. 2004 images of Titan were obtained showing interesting surface details (see Fig. 4.8). The Huygens probe was built by the European Space Agency. The Huygens probe separated on December 25, 2004 and successfully descended to Titan's surface on January 15, 2005. It landed on solid ground with no liquids in view. Although Cassini successfully relayed 350 of the pictures that it received from Huygens of its descent and landing site, a software error failed to turn on one of the Cassini receivers and caused the loss of the other 350 pictures (see Fig. 4.9).

Radar images obtained on July 21, 2006 appear to show lakes of liquid hydrocarbon (such as methane and ethane) in Titan's northern latitudes. This is the first discovery of currently-existing lakes anywhere besides Earth. The lakes range in size from about a kilometer to one which is one hundred kilometers across. Hydrocarbon seas on Titan's surface up to a size of $\sim$500 000 km^2 are described by Turtle

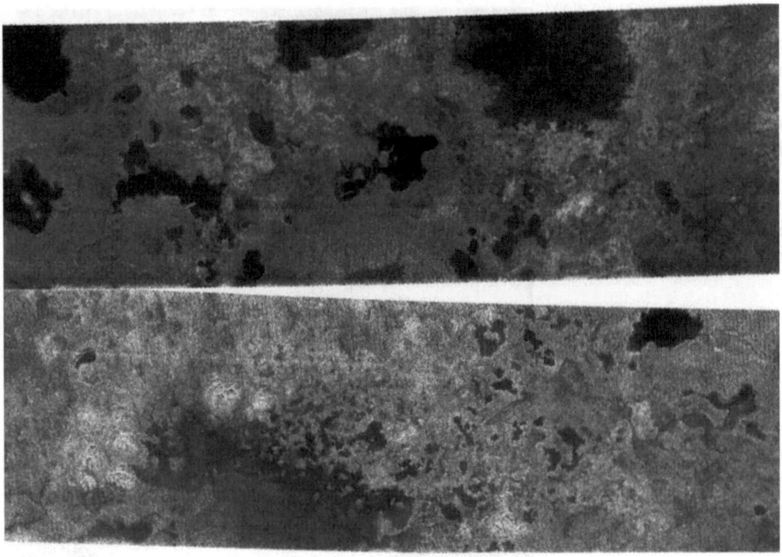

Fig. 4.9 The Cassini spacecraft, using its radar system, has discovered very strong evidence for hydrocarbon lakes on Titan. Dark patches, which resemble terrestrial lakes, seem to be sprinkled all over the high latitudes surrounding Titan's north pole. NASA/JPL/Space Science Institute

et al., 2009 [343]. The poles harbor liquid-hydrocarbon reservoirs, the extents of which differ from pole to pole and which may be coupled to seasonally varying circulation. The earthlike shoreline morphology of Titan's Ontario Lacus was studied by Wall et al., 2010 [351]. Evidence for active material transport and erosion was also found.

A review about the results from the Huygens probe was given by Lebreton et al., 2009 [197].

A review about what was known about Titan at the time of the launch of the Cassini missions was given by Taylor and Coustenis, 1998 [332].

Organic Haze in Titan's Atmosphere

A review of Titan's atmospheric phenomena was given by Hirtzig et al., 2009 [163].

Titans's atmosphere consists mainly of N_2 and CH_4. Photochemical reactions produce complex organics involving large amounts of hydrogen. In these processes EUV and VUV radiation are involved. Tholin (after the Greek word for muddy), is a heteropolymer formed by solar ultraviolet irradiation of simple organic compounds such as methane or ethane. Tholins do not form naturally on modern-day Earth, but are found in great abundance on the surface of icy bodies in the outer solar system. They usually have reddish-brown appearance. Tholins have also been detected in an extrasolar planetary system: HR 4796A which is a very young system (eight-million-year-old star) and its distance from the Earth is 220 light years. The detection was made with the Near-Infrared Camera and Multi-Object Spectrometer (NICMOS) aboard the Hubble Space Telescope (Debes et al., 2008 [101]). Köhler, Mann and Li, 2008 [188] showed that porous grains composed of common cosmic dust species (amorphous silicate, amorphous carbon, and water ice) also closely reproduce the observed reflectance spectrum, therefore the presence of complex organic materials in the HR 4796A disk is still not definitive.

For astrobiological studies it is important to note that tholins can act as an effective screen for ultraviolet radiation, protecting the surface from it. A wide variety of soil bacteria are able to use tholins as their sole source of carbon. It is thought tholins on Earth may have been the first microbial food for heterotrophic microorganisms before autotrophy evolved. Sekine et al., 2008 [304] studied the role and formation of tholins in Titan's atmosphere.

Neish et al., 2010 [246] investigated a tholin sample that had been hydrolyzed in a ammonia-water solution at 253 ± 1 K and 293 ± 1 K for 1 year. These four species have been assigned as the amino acids asparagine, aspartic acid, glutamine, and glutamic acid. So biologically relevant molecules can be created under conditions thought to be similar to those found in impact melt pools and cryolavas on Titan and the situation might be even similar to that of the primordial soup on early earth.

Water and Ice on Titan?

Water has been first detected with the Infrared Space Observatory (ISO) in the atmosphere of Titan. The Infrared Space Observatory (ISO) was launched by the ESA Ariane in November 1995 and was the first true orbiting infrared observatory for a detailed investigation of the universe in the infrared. It was equipped with a 60 cm diameter telescope which was cooled by superfluid liquid helium to temperatures of 2–4 K. Observations were made from 2.5 microns to 240 microns with a spatial resolution ranging from 1.5 arcseconds to 90 arcsec (at the longer wavelengths).

The detection of water in Titan's atmosphere in the thermal infrared (2.36–45.2 μm) was made from ISO observations in Jan. and Dec. 1997. Two pure rotational water lines were observed at 39.4 and 43.9 μm. The observed water abundance[7] is four times lower than in comets and suggests that Titan's atmosphere is not of cometary origin but rather formed by outgassing from the interior (Coustenis et al., 1998 [82]).

The Composite Infrared Spectrometer (CIRS)[8] on-board Cassini had an improved sensitivity and better resolution in the far IR than the ISO spectrometer. The measurements were in good agreement (Nixon et al., 2006 [249], Bjoraker et al., 2008 [33]).

---

[7] The amount detected is extremely low: 8×10^{-8} to 4×10^{-10}.

[8] The Composite Infrared Spectrometer (CIRS) consists of dual interferometers that measure infrared emission from atmospheres, rings, and surfaces over wavelengths from 7 to 1000 micrometers (1400 to 10 cm^{-1}) to determine their composition and temperatures. It consists of a 50-cm telescope.

Stellar occultations provide an opportunity to probe Titan's atmospheric structure and composition. On December 14, 2004 two stellar occultations were observed (the stars Spica and Shaula were occulted). The data for Titan's northern and southern hemisphere were compared. At the time when the occultations occurred it was late winter in the northern hemisphere. Water was found, however, in the Spica data below 575 kilometers, as predicted by earlier models. The restriction of the water ice clouds to the northern hemisphere, where it was late-winter, suggests a seasonal dependence on the distribution of Titan's water ice clouds (Larsen et al., 2006 [194]).

Correlation of features in overlaps of Cassini RADAR swaths allows determination of Titan's rotational pole position and spin rate. The initial analysis of these data shows Titan to be rotationally variable and to have an internal water ocean. The main argument for this is the following. The Earth's Length-of-Day changes by 1 ms over a year. Similarly, seasonal changes in Titan's atmospheric angular momentum would manifest themselves in a change in surface rotation rate. The change in rate is ten times higher, amounting to some hundreds of seconds, when the surface is decoupled from the interior by a water-ammonia ocean (Lorenz et al., 2007 [208]).

The Cassini Titan Radar Mapper observed on Titan several large features interpreted as cryovolcanic. Mitri et al., 2008 [232] propose a model of cryovolcanism that involves cracking at the base of the ice shell and formation of ammonia-water pockets in the ice. Due to varying buoyancy of these pockets, episodic geologic activity on Titan could be triggered. Barnes et al., 2007 [18] discussed evidence for cryovolcanism and tectonics from Cassini/VIMS[9] (Fig. 4.10).

Landforms on the surface of Titan are similar to viscous cryovolcanic flows and possibly eruptive domes. In an experiment, a 40% methanol-water mixture was tested to form such shurries (Zhong et al., 2009 [372]).

From the atmospheric composition we can infer the internal composition of a satellite. The temperature inside Titan probably reached values higher than 800 K or even higher than 1000 K, since the process of the chemical dissociation of ammonia was completely finished on this satellite and its atmosphere contains only molecular nitrogen (Dorofeeva and Ruskol, 2010 [109]).

Radar-bright channels are interpreted as riverbeds, where debris, likely shaped and transported by fluvial activity, have been deposited. Similar debris were observed in the landing site of the Huygens probe. The size of the icy peebles is larger than the radar wavelength (2.18 cm) (Cassini Radar Team, 2010 [333]).

Raulin, 2008 [274] describes Titan as a natural laboratory for prebiotic-like chemistry—an active organic chemistry, in the absence of permanent liquid water, on the surface.

4.2.3 Other Satellites of Saturn

The ring planet Saturn has 60 confirmed moons. These can be classified into

[9]Visual and Infrared Mapping Spectrometer.

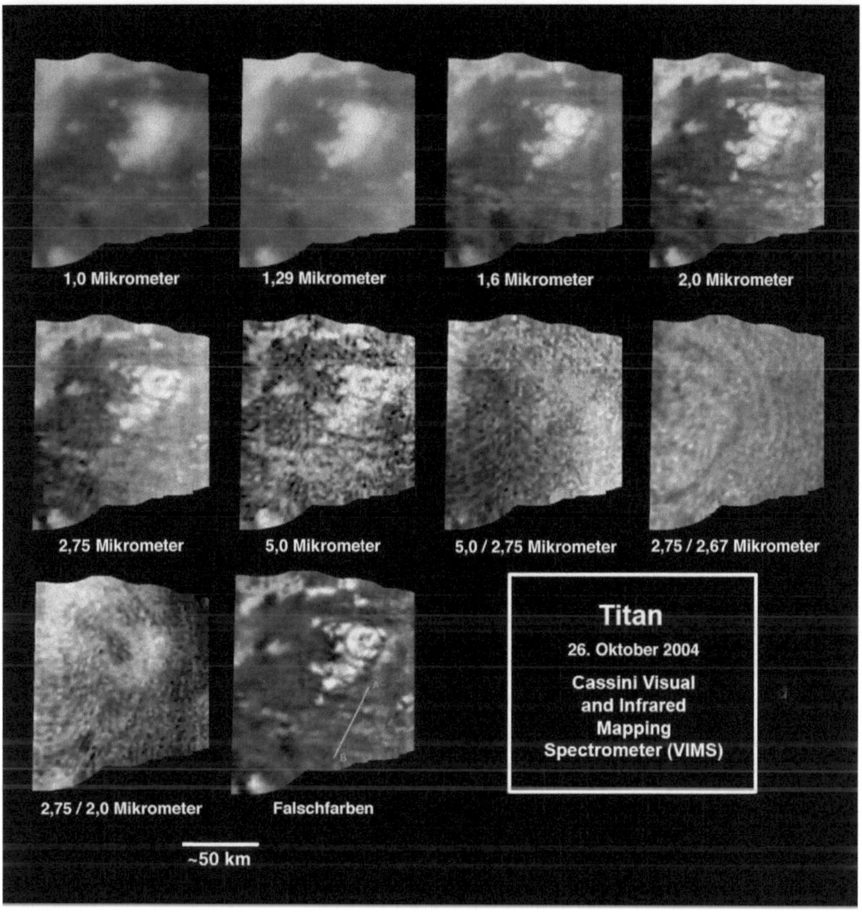

Fig. 4.10 The VIMS spectrometer on board Cassini photographed the 'snail shell', with a diameter of around 30 km, in high resolution and in several different wavelengths. These images give us a glimpse of the surface of Titan. Credit: Nature/Cassini, VIMS team

- 22 regular satellites; they have prograde orbits (i.e. they orbit the planet in the same sense as the planet rotates around its own axis). The seven largest are of spheroidal shape.
- 38 satellites move retrograde or prograde; their orbits have a greater inclination and they are shaped irregularly. These objects were captured by Saturn due to its high gravity.

Rhea

This moon is also designated as Saturn V. It was discovered in 1672 by Cassini who also discovered Tethys, Dione and Japetus. Rhea (Fig. 4.11) has a low density

Fig. 4.11 Rhea taken by the Cassini spacecraft January 17, 2006. The streaks are revealed to consist of ice cliffs similar to those of Dione. NASA/JPL/Space Science Institute

(1.233 g/cm^3) and is made of ~25% rocks and 75% water. The diameter is about 1500 km, the rotation period 4.5 days (synchronous rotation). The temperature on the surface of Rhea is 99 K on the dayside and 53 K on the nightside. The surface appears heavily cratered and shows also some bright wispy markings. There are two different areas on the surface which is 7.3×10^6 km^2: one which is covered by craters which are larger than 40 km in diameter and one (polar and equatorial regions) where the craters are smaller. This may be explained by some geologic process at the time of the formation of Rhea.

This satellite is studied by the Cassini mission and in 2008 it was found that it is surrounded by a ring system.[10]

The amount of water vapor that could be generated on the surface of Rhea by geologic activity can be estimated from high solar phase angle curves (phase angle curves for Rhea from ground based observations were only obtained up to a value of 70°) using data from VISM (Visual and Infrared Mapping Spectrometer on board Cassini). It was found that the maximum water vapor column density that could be supplied from Rhea range form 1.5×10^{14} to 1.9×10^{15} cm^{-2} indicating a very low level of active internal (endogenic) processes on Saturn's second largest moon (Pitman et al., 2008 [264]). Wagner et al., 2008 [349] presented a study of the geology of Rhea on the basis of the high-resolution images from the targeted flyby on Aug. 30, 2007. Castillo-Rogez, 2006 [61] modeled the interior of Rhea on the basis of observational constraints. Different models were discussed including high pressure layer ice (so called ice II) and varying extend of rock components from volatiles.

[10]Astronomy and Geophysics, Volume 49, Issue 2, p. 2.05.

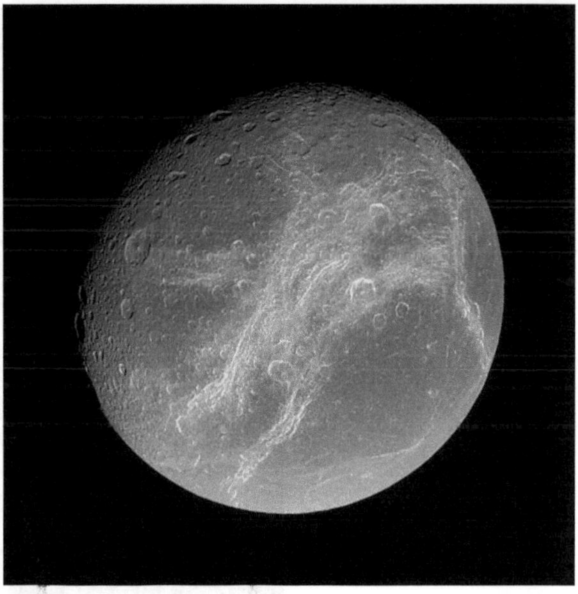

Fig. 4.12 Dione and its wispy terrain. NASA/JPL/Space Science Institute

The surface composition of Rhea from reflectance spectra between 0.2 to 3.6 μm using the UKIRT telescope on Mauna Kea and the CGS4 spectrometer, revealed strong bands of H_2O ice at 3.0 μm (Dalle Ore et al., 1999 [91]).

Dione

This satellite of Saturn is also designated as Saturn IV. Its semi major axis is 377 396 km, the orbital period is 2.73 d. The mean radius is 561 km and its mean density is 1.47 g/cm^3. Because of this relatively low density, this satellite must be mainly composed of water, about 45% seems to be silicate rocks which are mainly concentrated in its interior. Like Rhea its trailing and leading hemisphere are dissimilar. On its surface, wisps were found (see Fig. 4.12) which were identified by the flyby of the Cassini probe in December 2004 as bright ice cliffs created by tectonics.

Clark et al., 2008 [69] have reported on the compositional mapping of Dione with Cassini VIMS. The Dione VIMS data show a pattern of bombardment of fine, sub-0.5-μm diameter particles impacting the satellite from the trailing side direction and that the dark material on Dione is of external origin.

Tethys

This satellite is also designated as Saturn III. Like Rhea and Dione it is an icy body. The density of Tethys is only 0.97 g/cm^3. So it is almost entirely composed of ice. The surface appears as heavily cratered (see Fig. 4.13) and there are numerous cracks caused by faults in the ice. Its surface has a very high albedo which is a result

Fig. 4.13 Tethys, note the enormous (450 km diameter) impact structure named Odysseus. NASA/JPL/Space Science Institute

Fig. 4.14 Tethys, with Ithaca Chasma, a fracture valley that runs across Tethys. NASA/JPL/Space Science Institute

of sandblasting of particles from Saturn's E-ring. This ring is composed of small water-ice particles that are generated by Enceladus's south pole geysers. Tethys' diameter is 1080 km, the orbital period 1.88 days, the mean surface temperature 86 K: the semi major axis is 294 619 km.

On Tethys, a striking surface feature can be seen, a tremendous rift called Ithaca Chasma (see Fig. 4.14), which is 100 kilometers and runs nearly three-fourths of the way around the icy moon (Elder et al., 2007 [116]). One theory to explain the formation of this phenomenon is that Ithaca Chasma formed as Tethys' internal liquid water solidified, causing the moon to expand and cracking its surface to accommodate the extra volume within (because water ice is less dense than liquid water).

Tethys might have had a subsurface liquid ocean during its early history (Chen and Nimmo, 2008 [66]) when there was a 2:3 orbital resonance between Dione and Tethys. The resonance would have led to orbital eccentricity and tidal heating that may have warmed Tethys' interior enough to form the ocean. Subsequent freezing of the ocean after the moons escaped from the resonance may have generated the extensional stresses that created Ithaca Chasma.

An alternative hypothesis is that it was formed at the same time as the large crater Odysseus which is on the opposite side of the moon. When the impact that created Odysseus occurred, the shockwave may have traveled through Tethys and fractured the icy, brittle surface on the other side.

The surface must be very old, because it is heavily cratered.

Iapetus

Iapetus, is the third largest satellite of Saturn. Its mean radius is 735 km and the surface area is about 6.7×10^6 km^2. The density is only about 1 g/cm^3 so that it is composed by 80% of ice. The distance from Saturn is about ten times the distance Earth-Moon and its orbit is inclined by 15° with respect to Saturn's equator.

There is a strong difference in coloring between the two hemispheres of Iapetus: the leading hemisphere and sides are dark (albedo 0.03–0.05) with a slight reddish-brown coloring, while most of the trailing hemisphere and poles are bright (albedo 0.5–0.6, almost as bright as Europa). Hendrix and Hansen, 2008 [161] report on volatile migration processes in the boundary region of the bright and dark terrains.

A hint for the existence of two different hemispheres of Iapetus came from the measured variations of its apparent luminosity from Earth.[11] Iapetus has a very slow rotation period. In the paper of Castillo-Rogez et al., 2007 [62] it is described how its present 79.33 d rotation evolved from a 16 h rotation. Because of radioactive decay at its formation time, the internal temperatures raised and tidal dissipation could start, therefore Iapetus obtained its synchronous rotation several hundred million years after the satellite formed.

The satellite loses ice from sublimation, however this loss is different for the two different hemispheres (see Fig. 4.15, upper image). The sublimation rate for the bright region (called Roncevaux) is about 10 cm in one billion years, whereas this value is doubled for the dark Cassini hemisphere. Iapetus is locked in synchronous rotation. Therefore it has the warmest daytime temperature and the coldest nighttime

[11] Its apparent magnitude at mean distance of Saturn varies between 10.2 and 11.9 magnitudes.

Fig. 4.15 Iapetus, partly covered with ice and the equatorial ridge (*below*). Courtesy: NASA

temperature in the Saturnian system. The temperature is about 128 K in the Roncevaux region and 113 K in the Cassini region. This explains why ice preferentially sublimates in the Cassini region.

The dark material may be debris from other satellites (e.g. Phoebe) or micrometeorites. One further characteristic of Iapetus is an equatorial ridge (see Fig. 4.15, lower image) which is mainly seen in the Cassini hemisphere. This ridge extends about 1400 km and can rise up to 20 km above the local surface. The ridge could be icy material that welled up from beneath the surface and then solidified but there are also other theories. In any case, the ridge seems to be old because it is heavily cratered (Czechowski and Leliwa-Kopystyn'ski, 2008 [89]).

Enceladus

Enceladus is the sixth-largest moon of Saturn (also called Saturn II). Its diameter is about 500 km and its semi-major axis 238 000 km. Therefore, its orbital period is rather short, 1.37 days. The mean density is 1.6 g/cm^3. Again the surface can be

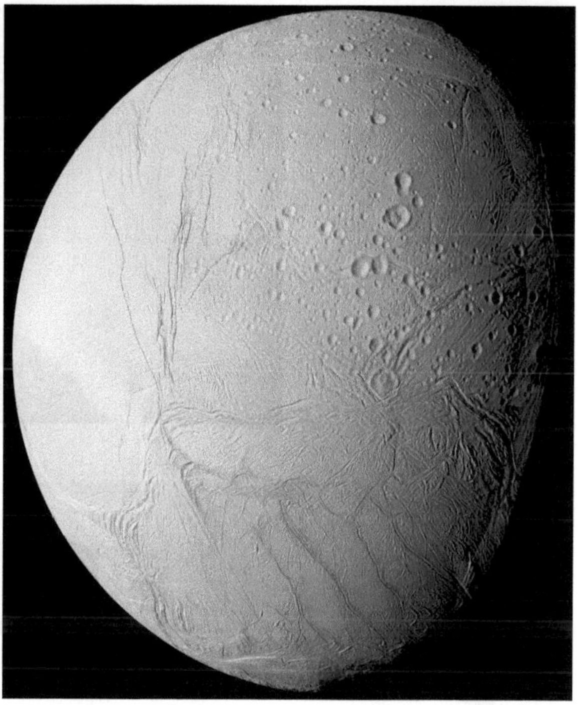

Fig. 4.16 The surface of Enceladus resembles that of Europa. Credit: NASA/JPL/Space Science Institute

divided into a heavily cratered mid- to high-northern latitude region, and a lightly cratered region closer to the equator. Fresh, clean ice dominates its surface (see Fig. 4.16) and the albedo is the highest of all known bodies in the solar system. Because it reflects so much sunlight, the mean surface temperature at noon only reaches −198°C.

During the flybys of Cassini in 2004, water vapor and complex hydrocarbons were detected which came from the geologically active South Polar region. This cryovolcanic plume provides the source for particles in Saturn's E ring. This active region was detected on July 14 2005 on a Cassini flyby. The measured thermal emission was between 3 to 7 GW and the temperature is about 145 K (Spencer et al., 2006 [316]).

Hodyss et al., 2008 [164] made a simulation by a mixture of gases at a temperature of 20 K with the in situ measured plume composition by the Cassini mission. With such an experiment they simulated the release of N_2, CH_4, CO_2, and H_2O from surface ices on Enceladus. Tobie, Cadek, and Sotin, 2008 [337] showed that only interior models with a liquid water layer at depth can explain the observed magnitude of dissipation rate and its particular location at the south pole. However, sustaining some liquid water at depth is impossible if heat is continuously emitted at a rate of 4–8 GW at the south pole and the current thermal emission is probably abnormally high at present time. Alternatively, continuous dissipation at a rate of 1–2 GW in the ice shell at the south pole should be sufficient to induce internal melting and it could sustain a layer of liquid water at depth over geologic timescales.

Fig. 4.17 Geysers on Enceladus. Courtesy: NASA/JPL/Space Science Institute

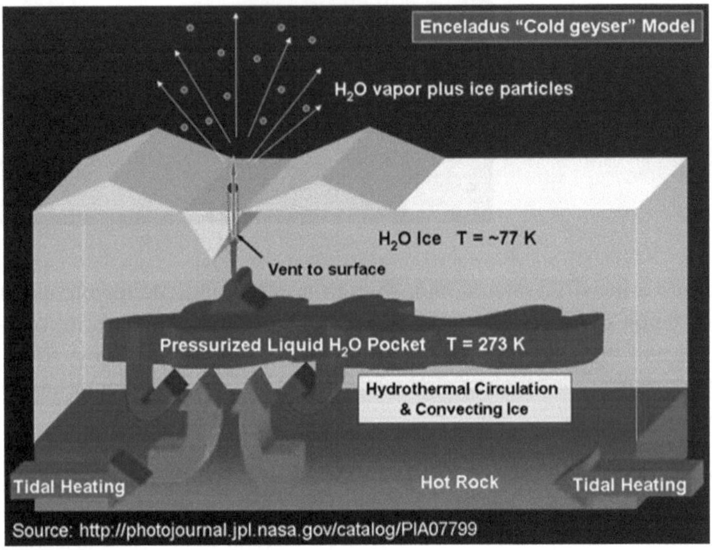

Fig. 4.18 Enceladus: cryovolcanism. Courtesy: NASA/JPL/Space Science Institute

The Cassini plasma spectrometer detected water-group pick-up ions in a toroidal region which is located along the orbit of Enceladus. The source of these ions is the cryovolcanic activity near the South pole. The density is estimated at about 5.2 cm^{-3} which corresponds to about 8% of the total ion intensity (Tokar et al., 2008 [338]).

McKinnon and Barr, 2008 [225] found that if there is ice convection an ocean-bearing Enceladus can adjust its oceanic thickness and composition so that satellite heat flows are steady state.

Ingersoll and Pankine, 2010 [169] studied the subsurface heat transport on Enceladus. Whether liquid water exists on Enceladus depends on the efficiency of subsurface heat transfer (see Fig. 4.18). Melting within 40 m of the surface occurs for a heat flux equal to 124 K black body radiative flux, however, there are estimates of regions on the surface with temperatures over 200 K (Abramov and Spencer, 2009 [2]). On images on the South Polar region the so called tiger stripes were found which are sources of intense IR radiation and plume activity (see Fig. 4.17). The plumes can be up to 500 km high. The existence of such a water plume was proven by the Cassini UVIS instrument (Ultraviolet Imaging Spectrograph). Enceladus's plume was observed when occulting the star γ Ori in 2005. This observation confirmed

- Location of the plume in the south polar region.
- Plume is dominated by water vapor (the line of sight column abundance 1.5×10^{16} cm^{-2}).
- Flux of water coming from Enceladus $\sim$150 kg/s.

On 24 October 2007, UVIS observed ζ Orionis (Alnitak) occulted by the plume (Hansen et al. 2006, 2008 [153, 154]). Water vapor transport may occur by (i) diffusion through pores in an icy matrix, (ii) hydrodynamic flow in open cracks. Approaching the surface, the molecules either escape as vapor or they condense close to the ice and transfer their latent heat to the ice.

Waite et al., 2009 [350] found that ammonia is present in the plume, along with various organic compounds, deuterium and, very probably, ^{40}Ar. The presence of ammonia provides strong evidence for the existence of at least some liquid water, given that temperatures in excess of 180 K have been measured near the fractures from which the jets emanate. Ammonia (together with methanol and salts) acts as an antifreeze that permits the existence of liquid water down to temperatures as low as 176 K. Schneider et al., 2010 [295] searched for sodium in Enceladus' water plumes. The lack of observable sodium in the vapor is consistent with a deep ocean, a freshwater reservoir, or ice. The plume particles are the most important source for particles in Saturn's E-ring. The observations are consistent with a subsurface ocean in contact with its rock core. There is no atomic sodium in the vapor however it is present in salts in Saturn's E ring (see also Spencer, 2009 [314]).

Spencer et al., 2009 [317] discussed Enceladus as an active cryovolcanic satellite. Plumes, both large and small, spray water ice out from many locations along the famed "tiger stripes" near Enceladus' south pole.

4.3 Satellites of Uranus and Neptune

4.3.1 The Satellites of Uranus

The two largest satellites of Uranus, Titania and Oberon, were already discovered in 1787 by W. Herschel. Later Ariel, Umbriel and Miranda were found. The remaining ones were discovered by the Voyager 2 space probe in 1986. The smallest moons move on strongly perturbed orbits and collisions are likely to happen, e.g. Desdemona may collide either with Cressida or Juliet within the next 100 million years.

Table 4.3 Some important parameters of some satellites of Uranus

Name	Mean diameter (km)	Mass ($\times 10^{16}$ kg)	Semi-major axis (km)	Orbital period (day)
Miranda	471	6600	129 390	1.41
Ariel	1157	135 000	191 020	2.52
Umbriel	1169	117 000	266 300	4.13
Titania	1577	353 000	435 910	8.71
Oberon	1522	301 000	583 520	13.46
Juliet	107	~82	64 360	0.49
Portia	141	~170	66 097	0.51
Belinda	100	~49	75 255	0.62
Puck	162	2 ~ 90	86 004	0.76

Up to now 27 satellites of Uranus are known. The properties of Uranus' satellites with a diameter >100 km are listed in Table 4.3.

Titania (Uranus III, Fig. 4.19) is the largest moon of Uranus and the eight-largest moon in the solar system. Its mean density is 1.72 g/cm^3 and its radius is 789 km. The albedo is 0.27 and the temperature about 60 K. Like the Vallis Marineris on Mars or the Ithaca chasma on Tethys, a huge canyon is a dominant surface phenomenon. It is assumed that Titania is composed of 50% water ice, 30% silicate rock and 20% organic compounds including methane. Grundy et al., 2006 [147] made spectral observations at 0.8 to 2.4 μm. The spectral bands observed near 2 μm indicate the presence of CO_2 ice on the surface of Umbriel and Titania. In the spectra of Oberon, no signs of CO_2 absorption were found. The longitudinal distribution of CO_2 seems to be not uniform on Ariel, Umbriel and Titania—it is more abundant on their trailing hemispheres. On the leading atmospheres of Ariel, Umbriel and Titania deeper H_2O bands were found. Titania occulted a bright star (HIP 106829) on September 8, 2001. This occultation was observed with telescopes of different sizes (from 20 cm to 100 cm). The main aim was to find out, wether Titania has an atmosphere. The upper level for such an atmosphere was found to be at 0.1 microbar (Sicardi et al., 2001 [310]).

The cratering history on the moons of Uranus was investigated by Plescia, 1987 [265]. The surfaces of Oberon and Umbriel are interpreted to be the oldest, that of Titania intermediate and those of Ariel and Miranda the youngest.

Umbriel (Fig. 4.20) is the darkest Uranian satellite, its mean density is 1.4 g/cm^3. Its surface is more heavily cratered than that of Ariel and Titania, so its is older and there are no signs of geologic activity. Near its equator, a crater like bright structure, called Wunda is found. The crater Skynd possesses a huge bright central peak but no rim.

Ariel (Uranus I, Fig. 4.22) has a mean density of 1.67 g/cm^3 and is composed of 70% of ices (water ice, carbon dioxide ice and methane ice). On its surface deposits of fresh frost are visible. The largest crater found was named Yangoor (78 km diameter). Also a network of faults and canyons is seen. The origin of these pat-

Fig. 4.19 High-resolution color composite of Titania (Voyager 2 images taken Jan. 24, 1986, from a distance of 500 000 km). The picture shows details about 9 km. The large, trenchlike feature near the terminator (day-night boundary) at middle right suggests at least one episode of tectonic activity. Credit: NASA/JPL/Space Science Institute

Fig. 4.20 The darkest satellite of Uranus, Umbriel (Voyager 2, Jan. 24, 1986). The albedo is 0.16 which is comparable to the lunar highlands (Terrae). Near the satellite's equator a ring is seen (diameter 140 km). This might be a frost deposit. Credit: NASA/JPL/Space Science Institute

terns was simulated by Nyffenegger, Davis and Consolmagno, 1997 [251]. Spectra reveal contrasts between Ariel's leading and trailing hemispheres, with the leading hemisphere presenting deeper H_2O ice absorption bands. The observed dichotomy is comparable to leading-trailing spectral asymmetries observed among jovian and

Fig. 4.21 The IRTF, Infrared Telescope Facility. Credit: NASA, Institute of Astronomy, Univ. of Hawaii

Saturnian icy satellites. The trailing hemisphere spectrum exhibits three narrow CO_2 ice absorption bands near 2 μm (Grundy, Young, L.A. and Young, E.F., 2003 [146]). These spectra were obtained with the IRTF (Infrared Telescope Facility, Fig. 4.21), a 3 meter telescope optimized for IR observations at Hawaii.

Oberon is the second largest satellite of Uranus and the tenth-largest in the solar system. It is composed 50% of water ice, 30% of silicate rock and 20% of methane-carbon/nitrogen compounds. There is no evidence for activity on its surface which is heavily cratered and therefore quite old. Some large faults are found across the southern hemisphere.

The maintenance of oceans in smaller bodies depends on a balance of internal heat generation and thermal isolation by ice or other insulating material. If the surface of moons is covered by impure ice, the thermal conductivity becomes reduced which makes the likelihood of oceans higher. Tidal heating is important, in the case of the Jupiter satellite Europa it amounts to over 40 times its internal radiogenic heating. The investigations of England, 2003 [122] show that subsurface oceans are likely for Titania, Rhea, Oberon and Iapetus but they will occur at a depth of 300 km under insulating ice layers. Surface features on these objects should provide a hint for the existence of such deep lying oceans.

Fig. 4.22 Ariel's extensive faulting has occurred as a result of expansion and stretching of that satellite's crust. The largest fault valleys, near the terminator at right, as well as a smooth region near the center of this image, have been partly filled with deposits that are younger and less heavily cratered than the pitted terrain. Credit: NASA/JPL/Space Science Institute

Using the Faint Object Spectrograph (FOS) on the Hubble Space Telescope (HST), ultraviolet spectra were obtained for the Uranian satellites Ariel, Titania, and Oberon. The absorption found at 280 nm can be explained by OH which is a byproduct from photolysis of H_2O (Roush et al., 1997 [283]).

4.3.2 The Satellites of Neptune

Neptune has thirteen known moons (see also Miner and Wessen, 2002 [231]). The largest satellite is Triton (Fig. 4.23) which is on retrograde orbit around Neptune. Only six moons have diameters larger than 100 km.

Some important parameters for satellites of Neptune larger than about 100 km are given in Table 4.4.

Triton is the largest moon of Neptune and was detected in 1846. It is the seventh-largest moon in the solar system and the only large moon with retrograde orbit.[12] It is similar in composition to the dwarf planet Pluto. Triton is thought to have been captured from the Kuiper belt. Its mean density is 2.06 g/cm^3. It is composed of approximately 15–35% water ice. Triton is geologically active, its crust is dotted with geysers which erupt nitrogen. Therefore, Triton also has a tenuous nitrogen

[12]It moves around the planet opposite to the planet's rotation direction.

Fig. 4.23 Triton, the largest moon of Neptune is slightly smaller than our Moon. Credit: NASA/JPL/Space Science Institute

Table 4.4 Some important parameters of some satellites of Neptune

Name	Mean diameter (km)	Mass ($\times 10^{16}$ kg)	Semi-major axis (km)	Orbital period (day)
Despina	180 × 150 × 130	19	48 227	0.29
Galatea	204 × 184 × 144	213	61 953	0.43
Larissa	216 × 204 × 164	420	73 548	0.56
Proteus	436 × 416 × 402	4 400	117 647	1.2
Triton	2707	2 140 000	354 800	−5.87
Nereid	340	2200	5 513 400	360.14

atmosphere. There are predictions that Triton will collide with Neptune in about 3.6 billion years, maybe forming a new ring of Neptune. Triton's surface area is 23 million km^2. Triton may be differentiated like Earth into a solid core, a mantle and a crust. Water is found in the mantle and because of heat production by radioactive decay, there may be a subsurface ocean. The surface temperature is about 40 K. The volcanic activity is thought to be driven by seasonal heating from the Sun, unlike the tidal heating responsible for the volcanoes of Io. Evidence for the geologic activity comes also from the fact that on Triton's surface only few craters are observed. The Cantaloupe Terrain, in the western hemisphere of Triton, consists is mostly dirty frozen water. The surface age of Triton was recently reviewed by Schenk, and Zahnle, 2007 [294].

Cruikshank et al., 2000, [87] discussed water ice on Triton's surface by the analysis of the broad absorption bands in the near infrared at 1.55 and 2.04 μm. Crystalline

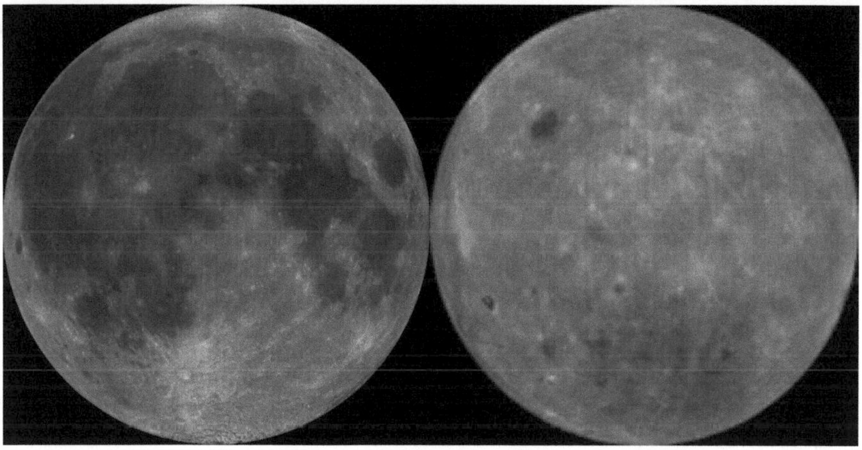

Fig. 4.24 Moon: near side (*left*) and far side (*right*). Note the absence of the dark appearing big plains (maria) on the far side. Image credit: NASA, Clementine mission

water ice has a distinctive absorption signature at 1.6 μm. However their results were not unambiguous whether Triton's water ice is crystalline or amorphous.

4.4 The Earth Moon

The Earth's Moon is the fifth largest satellite in the solar system. Its distance from the Earth is only 30 times the diameter of the Earth, the orbit is elliptical. The nearest distance to Earth (perigee) is 363 104 km, the largest distance (apogee) is 405 696 km. The semi major-axis is 384 399 km. The orbital period is 27 d 7 h 43.1 min which exactly corresponds to its rotational period. The orbit of the Moon is inclined to the plane of the ecliptic by about 5.1°. The mean radius of the Moon is 1737.1 km which is 0.273 that of the Earth's radius. The surface area is about 37×10^6 km^2. The density is relatively high and corresponds more to that of terrestrial planets and is 3.34 g cm^{-3}. Our Moon is a relatively dark body with a surface albedo of 0.12.

The Moon is in geosynchronous rotation keeping nearly always the same face towards Earth. In fact because of the so called libration effects, we can see about 59% of its surface from Earth. The side that face towards Earth is called the near side, the other hemisphere is called the far side (sometimes also the dark side; however this is not correct since also the far side is illuminated by the sun as is the near side). However, the far side was unknown before it was first photographed by the Soviet probe Luna 3 in 1959. The two sides are different (see Fig. 4.24), the far side is almost completely absent of the big plains that are called maria.

The lunar terrains are generally divided into bright highlands (terrae) which are heavily cratered and dark plains (maria) where few craters are found. The maria constitute 16% of the lunar surface but as it has been pointed out already, on the near side, the maria cover about 30% of the lunar surface. The number of impact craters is enormous, up to a size of 1 km about half a million craters are found. The impact craters per unit area accumulate with time, therefore the more heavily cratered the surface, the older it has to be. There are enormous impact basins like the Mare Imbrium. The surface of the Moon is covered by Regolith. Its thickness is between 3 and 5 m over the maria and 10–20 m over the highlands.

4.4.1 Water on the Moon?

In 1961 Watson, Murray and Brown [354] published a paper on the possible presence of ice on the Moon in craters that permanently are shadowed. Arnold, 1979 [10] mentioned four possible sources of lunar ice:

- solar wind reduction of Fe in the regolith,
- H_2O-containing meteoroids,
- cometary impact, and
- degassing of the interior which is the least certain.

They estimate lunar ice at a mass between 10^{13} and 10^{14} kg.

The two destructive mechanisms for lunar water ice were already mentioned briefly:

- sputtering or decomposition by solar wind particles,
- photodissociation of H_2O molecules by absorbed solar UV radiation.

With the Arecibo radar mapping also the search for ice on the lunar poles was performed (Stacy, Campbell and Ford, 1997 [323]).

Because of the large number of impacts especially from comets, water ice must have been deposited there. This H_2O is split by the incoming solar UV radiation (there is of course no protection against UV radiation because of the lack of an atmosphere) and the resulting hydrogen and oxygen quickly escapes into space because of the Moon's low gravity. The moon's rotation axis is slightly inclined (by about 1.5°). Therefore, some deep craters near the poles are never exposed to solar radiation. From computer simulations it became evident that an area of up to 14 000 km^2 might be in permanent shadow. The presence of water ice on the Moon is of great interest for future space missions. Water could be used for permanent lunar stations suggesting tremendous costs savings since transport of water through space is extremely expensive. The Clementine mission[13] was launched on January, 1995. The main task was to map the moon. It was an innovative mission in many senses: it took only 22 months from its conceptual design to launch and it was a cheap mission (costs of about 80 million USD). All other space missions before were significantly more expensive and it took more time from the conceptual phase to real launch. Between February 26 and April 22, 1994, Clementine was able to

[13]Also known as Deep Space Program Science Experiment, DSPSE.

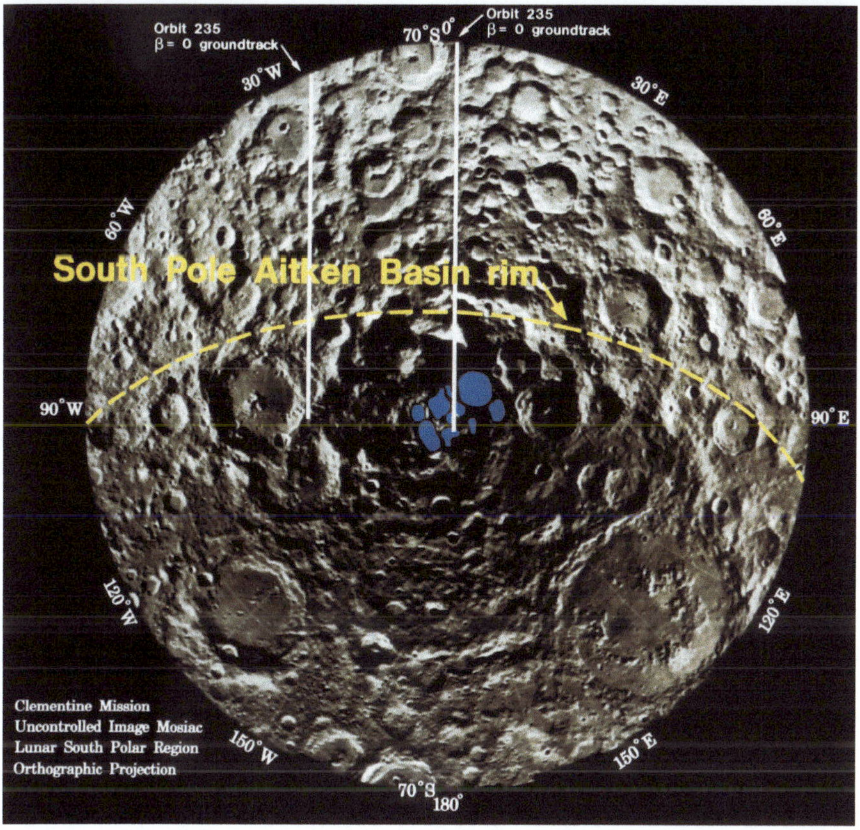

Fig. 4.25 Moon: Southpole with ice (marked in *blue*) Image credit: NASA, Clementine mission

deliver more than 1.8 million digital images of the moon. One of the major scientific discoveries was the finding of the existence of water ice in some lunar craters. Radar data indicated the presence of ice in some craters on the lunar South pole (see Fig. 4.25). The lunar surface (38×10^6 km^2) was mapped at eleven different wavelength from the UV (415 nm) to the near IR (2800 nm). The planned close flyby on the near-Earth asteroid Geographos was not possible due to a malfunction which left the spacecraft spinning at 80 revolutions per s.

Optical and near-IR signatures of water ice on the Moon's surface were investigated by McConnochie, 2002 [223]. At a wavelength of 1.5 μm they found signatures of water in some polar regions, however these signatures did not show up at 2 μm.

Regions on the lunar surface, that are permanently shaded are only found on the South pole (total area about 15 000 km^2). For detection the lunar surface composition, a radio transmitter was used. Rocky surfaces scatter radio waves randomly, while icy surfaces reflect radio waves coherently like a bicycle reflector.

The Lunar Prospector, a NASA Discovery mission, was launched into lunar orbit in January 1998. Included on Lunar Prospector is an experiment called the Neutron

Spectrometer. This experiment is designed to detect minute amounts of water ice at a level of less than 0.01%. The main interest of this mission were the lunar poles in order to search for water ice.

The neutron spectrometer was also used for Mars missions. Neutrons can be detected in three energy bands: thermal, epithermal and fast. The free neutrons are produced by collisions between cosmic ray particles and planetary matter. Hydrogen is a very good moderator for neutrons. The detector is sensitive to the presence of hydrogen on the surface (to a depth of about one meter). Large concentrations of hydrogen can be interpreted by the presence of water (either in liquid or in ice form). Data of the Lunar Prospector mission Neutron Spectrometer show a distinctive 4.6 percent signature over the north polar region and a 3.0 percent signature over the south, a strong indication that water is present in both these areas. The data show that water may occur in localized areas over the north and south pole (1850 km^2). The estimated total mass of water ice is about 6.6×10^9 tons. The results of the Prospector mission showed a much greater water ice reservoir in the lunar north pole than on the south pole (around the Aitken Basin). For comparison: on Earth, the column of permanent snow and ice on Greenland is about 3.0×10^6 km^3 and for the whole northern hemisphere 11×10^6 km^3.

Saal et al., 2008 [286] discussed volatile content of lunar volcanic glasses and the presence of water in the Moon's interior. Their conclusion from measurements of contents of the most primitive basalts in the Moon—the lunar volcanic glasses—is that water (and other volatiles like CO_2) must be considered in models constraining the Moon's formation and its thermal and chemical evolution. Chaudisson, 2008 [65] also deduced from analyzes of lunar volcanic glasses that they are rich in volatile elements and water. If the moon was formed by a huge impact on the early earth by a planet sized object this would have also consequences for models of early Earth formation.

The migration and accumulation of water into the subsurface was investigated by Schorghofer and Taylor, 2007 [296]. A few hundred ppm of H_2O could be deposited there being absorbed. The migration and deposit of water to deeper layers is important to protect its escape by two processes:

- sputtering
- photolysis by Lα.

The diurnal temperature fluctuations cause surface ice deposits to migrate through the lunar regolith.

The inside of Shackleton Crater at the lunar south pole is permanently shadowed; it has been inferred to hold water-ice deposits. With the Terrain Camera (TC), a 10-meter-resolution stereo camera onboard the Selenological and Engineering Explorer (SELENE, launched Dec. 2007) spacecraft this region was scanned, the temperature there is about 90 K. However, no signs of ice derived from the albedo measurements were found. Water-ice may be mixed with soil components however (Haruyama et al., 2008 [159]). Campell et al., 2006 [54] also found no evidence for thick deposits of ice at the lunar south pole.

Observations made with NASA's Moon Mineralogy Mapper M^3 (see Fig. 4.26) aboard an Indian satellite mission showed evidence for water on the Moon which

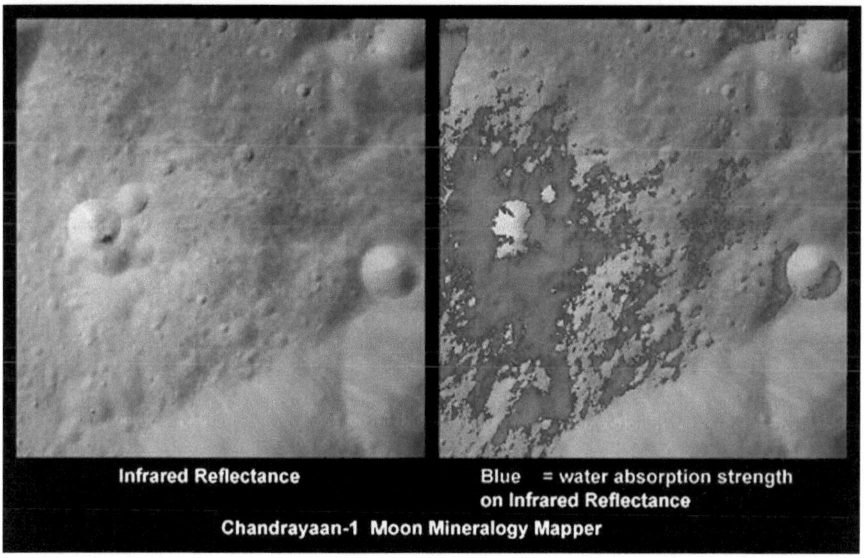

Fig. 4.26 Moon: A very young lunar crater as viewed by NASA's Moon Mineralogy Mapper. *On the right*, the distribution of water-rich minerals is shown in false-color blue. The observations were made by NASA's Moon Mineralogy Mapper, or M3 ("M-cubed"), aboard the Indian Space Research Organization's Chandrayaan-1 spacecraft (launched Oct. 2008)

means molecules of water and hydroxyl that interact with molecules of rock and dust specifically in the top millimeters (see Pieters et al., 2009 [261]) of the Moon's surface. Such regions were also found on the sunlit side of the Moon and the water signature appears stronger at higher latitudes.

Thus there is still some controversy about the existence of ice on the Moon.

Chapter 5
Water on Small Solar System Bodies

The group of small solar system bodies, SSSBs, comprises asteroids, comets and smaller objects. The most part of the smaller objects is concentrated in belts. The orbits of the objects of the Main Asteroid Belt between Mars and Jupiter and the Kuiper Belt outside the orbit of Neptune are concentrated near the ecliptic (plane of the solar system), the objects of the Oort cloud are distributed spherically and extend as far as to 50 000 AU from the Sun.

Comets and asteroids consist mainly of unchanged remnant debris from the solar system formation process some 4.6 billion years ago. This and the fact that many impacts of these objects on terrestrial planets especially in the early solar system occurred explain the interest to study them.

5.1 Clouds of Particles

5.1.1 The Kuiper Belt

General Properties of KBOs

The Kuiper belt extends from about the orbit of Neptune (30 AU) to about 55 AU and even more. The first Kuiper Belt Object, KBO, was discovered in 1992, presently several thousand KBOs are known. Their total number is estimated over 70 000 for objects over 100 km in diameter. It can be compared with the asteroid belt which occurs between the orbits of Mars and Jupiter but there are two important differences:

- Size: 20 times as wide.
- Mass: 20–200 times as massive.
- Composition: asteroid belt is mainly made of rocky matter; Kuiper belt objects are mainly composed of ice, frozen volatiles, methane, ammonia and water.

A. Hanslmeier, *Water in the Universe,* Astrophysics and Space Science Library 368,
DOI 10.1007/978-90-481-9984-6_5,

Orbits of KBOs

The orbits of the KBOs are nearly in the plane of the ecliptic. Scattered disk objects have high eccentric orbits (one example is Eris, $e = 0.44$) with aphelia up to 100 AU. The origin of these scattered objects was the outward migration at the formation of planet Neptune (Trujillo, Jewitt, and Luu, 1999 [341]). The Centaurs are on orbits among the gas giants and Neptune's moon Triton is believed a captured KBO. The largest known member of the KBO is the dwarf planet Pluto. A recent review about the origin of the Kuiper Belt object was given by Gomes, 2009 [144].

From the observations two populations of KBOs are discussed:

- Hot Population: inclinations 0.2–0.6 radians.
- Cold Population: inclination about 0.1 radians.

The existence of two different populations follows not only from the difference in their inclination but also from the fact they differ in color. The authors conclude that the hot population was formed by a single close stellar encounter at a distance of 80 000 AU. These caused an increase in the eccentricities of the hot population objects and their perihelia moved to 35 AU. Because of the outward migration of planet Neptune (see e.g. Levison and Morbidelli, 2003 [200]) most of them were removed, less then 10% remained in stable orbits. This is consistent with the observations of the present distribution. Neptune's migration was also studied by an N-body simulation (Hahn, Malhotra, 2005 [152]). Neptune's migration can explain Neptune's 5:2 resonance and other resonances.[1] Also Centaurs can be produced by such simulations. It is also estimated that the total number of KBOs having radii >50 km orbiting interior to Neptune's 2:1 resonance is $N \sim 1.7 \times 10^5$.

Formation and collisional evolution of Kuiper Belt Objects was investigated by Kenyon et al., 2008 [182].

We just mention that there is a confusion about naming. The term Trans Neptunian Object (TNO) is used for objects at the outer edge of the Solar system (see Fig. 5.1); here, the KBO are only a subset. Also the term Edgeworth-Kuiper belt is used, and the objects are called EKOs. A review on Trans Neptunian Objects was given by Schulz, 2002 [299]. Many objects are found near the 1:2 resonance with Neptune (while Neptune completes two revolutions about the Sun, the KBO completes only one).

Physical Parameters of KBOs

The expected surface temperature of the KBOs is about 50 K. Information about their composition can be gained by measuring their colors. The colors found range from neutral grey to deep red. This implies surfaces composed of dirty ices to hydrocarbons. This inhomogeneity in surface composition is difficult to explain. It was first assumed that all KBOs must be uniformly dark objects having lost most of their volatile ices from effects by cosmic rays.

---

[1] E.g. Pluto is in 2:3 resonance to Neptune.

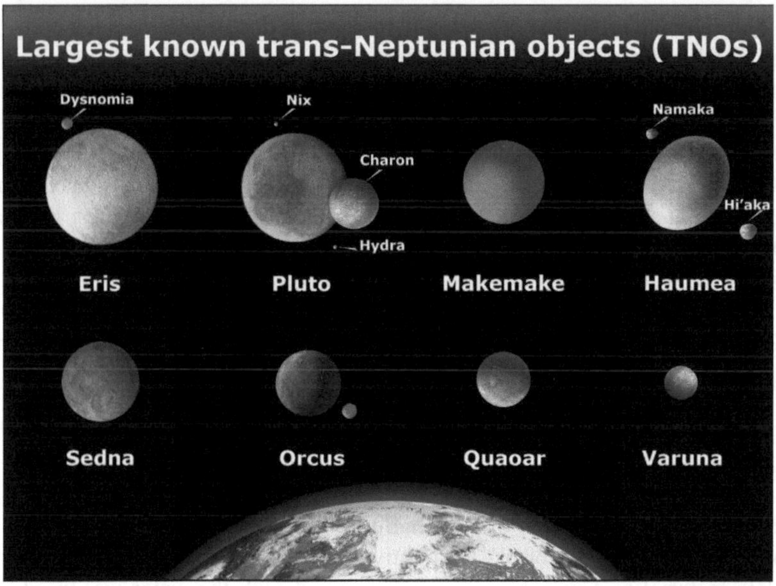

Fig. 5.1 Comparison of some known Trans Neptunian Objects, TNOs. Courtesy: A. Feild (Space Telescope Science Institute)

The already mentioned NICMOS instrument on the Hubble Space telescope was used to measure a sample of KBOs (84 objects were selected) photometrically and first results were presented by Noll et al., 2005 [250]. Stellar occultations provide another effective means to study these objects (see Roques and Doressoundiram, 2005 [281]). These observations were made with the WHT[2] in La Palma and with the Very Large Telescope of ESO. The influence of a stellar fly-by encounter on the KBOs was investigated by Kobayashi, Ida, and Tanaka, 2005 [187].

The upper limit for the mass of all KBOs is about 1/10 of the Earth's mass. From the accretion models that explain the formation of the objects in the early solar system it has to be expected that the number of objects of a population is inversely proportional to the diameter, D:

$$\frac{dN}{dD} \sim D^{-q}, \qquad q \sim 4 \tag{5.1}$$

Thus, if we find one object with the diameter of 1000 km, we can expect thousand objects with a diameter of 100 km.

Light curve variations may be caused by albedo variations (due to distinct surface features) or irregular shape. With the Advanced Camera for Surveys on board the Hubble Space telescope three small KBOs were analyzed for brightness variations. E.g. the object 2003 BF_{91} has a diameter of 20 km (under the assumption of an albedo of 0.1) and the amplitude of its light curve is 1.09 mag and the period

[2]William Herschel telescope, diameter: 4.2 m.

is 9.1 hr. The interpretation of such a light curve is that of an elongated body. Two other small bodies (2003 BG_{91}, diameter 31 km and 2003 BH_{91}, diameter 18 km) exhibited no detectable light curve variation. The object 2000 FV_{53} (116 km diameter) has a nonsinusoidal variation with an amplitude of 0.07 mag (period 7.5 hr). This is most likely a nearly spherical object (Trilling and Bernstein, 2006 [340]).

A Pluto sized object in the Kuiper belt is 2003 EL_{63}. Its apparent visual magnitude is 17.5 at a distance of 51 AU from the Sun. It is the third brightest KBO after Pluto and 2005 FY_9. The rotation period is 3.9 hr and the amplitude of its lightcurve is 0.28 mag which is a very fast rotation for a body larger than 100 km. The data are consistent with a diameter ranging from 1960 to 2500 km and a mean density of 2.6–3.3 $\mathrm{g\,cm^{-3}}$ and a visual albedo greater than 0.6 indicating an icy surface (Rabinowitz et al., 2006 [270]).

Water Ice on KBOs

The 1.40–2.45 μm spectrum of Kuiper Belt object 1996 TO_{66} was measured suggesting the presence of a black- to slightly blue-colored, spectrally featureless particulate material as a minority component mixed with the water ice. In addition, there is evidence that the intensity of the water bands in the spectrum of 1996 TO_{66} varies with rotational phase, suggesting a "patchy" surface (Brown et al., 1999 [48]).

Water ice was detected in the Centaur 2060 Chiron by Luu, Jewitt, and Trujillo, 2000 [209]. This object shows a cometary like activity which seems to decrease.

General remarks about water ice in the KBOs and the presence of organics there can be found in the contribution of de Bergh, 2004 [99]. As it was pointed out, the higher ice is a result of refreezing of water flowing upwards. Then there is a layer of amorphous ice and crystalline ice, the ratio between them continuously decreases outwards. The surface is a 20 km thick layer (22% of the volume) which has preserved its initial composition.

The large Kuiper belt object (50 000) Quaoar reveals the presence of crystalline water ice and ammonia hydrate. Crystallinity indicates that the ice has been heated to at least 110 K. Both ammonia hydrate and crystalline water ice should be destroyed by energetic particle irradiation on a timescale of about 10^7 yr therefore Quaoar has been recently resurfaced, either by impact exposure of previously buried (shielded) ices or by cryovolcanic outgassing, or by a combination of these processes (Jewitt and Luu, 2004 [171]).

Methane and ethane on the bright KBO 2005 FY_9 was reported by Brown et al., 2007 [47]. The spectrum from 0.34 to 2.5 μm is dominated by the red coloring of many outer solar system objects in the optical wavelength regime and by absorption due to methane in the near-infrared. The solid methane absorption lines are significantly broader on 2005 FY9 than on any other solar system body, indicating long optical path lengths through the methane. It is assumed that methane occurs in grains of about 1 cm size on its surface. Ethane is probably produced by a photolysis of methane.

Infrared H_2O band profiles when water is diluted in solid matrices dominated by methane, carbon monoxide, and/or molecular nitrogen that are expected on the surfaces of KBOs, were studied by Fulvio et al., 2010 [135]. Solid samples were analysed by infrared (1.0–16.0 μm) transmission and reflectance spectroscopy at different temperatures (15–150 K), before and after ion irradiation with 200 keV protons. They pointed out several features such as two narrow bands around 1.89 μm and 1.39 μm instead of the well known pure water ice bands at 2 μm and 1.5 μm, respectively.

The evolution of icy bodies of the Solar System can be modeled by adaptive grid numerical techniques. Such a code was used for evolutionary calculations spanning 4.6×10 yr of a growing body made of ice and rock, starting with a 10 km radius seed and ending with an object 250 km in radius (Prialnik and Merk, 2008 [267]). They adapted their code to two different objects (i) a typical KBO and (ii) Enceladus. In all cases the melting point of ice is reached in a central core. Evaporation and flow of water and vapor gradually remove the water from the core and the final (present) structure is differentiated, with a rocky, highly porous core of 80 km radius (and up to 160 km for very low conductivities). Outside the core, due to refreezing of water and vapor, a compact, ice-rich layer forms, a few tens of km thick (except in the case of very high conductivity). If the ice is initially amorphous, as expected in the Kuiper belt, the amorphous ice is preserved in an outer layer about 20 km thick. Let us give some details for these simulations: First of all the models take into account the accretion phase in the first 100 Myr of solar system evolution. Heating by this process is neglected. During the accretion phase the body grows in size from its initial radius of 10 km to a final radius of 250 km. The half lifetime of ^{26}Al is 0.72 Myr which is short in comparison with the accretion phase. Therefore most of the radiogenic heating has occurred when the body has only doubled in radius, i.e. its volume increased by a factor of 10. The central temperature reaches a peak after about 2 Myr of growth and then starts to decrease gradually. The radioisotope ^{40}K is longer lived, the half-life is 1.25×10^9 yr. however the energy is released at a much lower rate. Due to accretion and resulting growth of the object, the heat diffusion timescale of the growing body increases the radioactive heating timescale, the central temperature starts to rise again. This happens when about one third of the mass has been accreted and the radius exceeded 175 km. A third but lower peak is reached when the body has reached its final size of 250 km. Concerning ice, there occurs a transition at about 120 K. At that temperature, the amorphous ice starts crystallization. During time, this crystallization front moves outwards. Since heating in the core continues, the core starts to sublimate and the water vapor flows outward. However most of this vapor freezes again when moving outwards. By these processes however, the core becomes depleted of ice. At the end of the model simulation (4.6 billion yr) only a small temperature gradient between the core and the surface remains: at the core, the temperature is 60 K, at the surface 40 K. A porous ice depleted rocky core (radius 80 km) is surrounded by a thin layer enriched in ice by about 10% with respect to the initial composition.

Satellites of KBOs

KBOs even have satellites, some of them might be captured objects.

The spectrum of the brightest satellite of the large KBO 2003 EL_{61} showed absorption features at 1.5 and 2.0 μm which indicates water ice on its surface. The observations were made with the Keck-Telescope using the NIRC facility. The satellite was seen from Earth at a distance of only 1.4 arcsec.[3] Since these absorption features appeared much stronger than on other KBOs, it can be assumed that the satellite was formed by impact and not by capture (Barkume, Brown, and Schaller, 2006 [16]), Brown et al., 2007 [47]). Another example of KBO binary observations is object 2003 QY_{90}. The rotation periods and amplitudes of the lightcurves for the two components are: 3.4 and 7.1 h, and 0.34 and 0.90 mag. The orbital period of the system is estimated at 300 000 days and a minimum semi-major axis is given 13 092 km (Kern and Elliot, 2006 [183]).

5.1.2 The Oort Cloud

Objects of the Oort Cloud, Parameters

Already in 1932 E. Öpik assumed that long-period comets originate in a cloud at very large distance from the inner solar system. Since the long-period comets have random orbits, such a cloud must be distributed spherically around the solar system. Oort, 1950 [255] postulated its existence from the volatile composition of comets and their orbits (see Fig. 5.2). He found a peak in the numbers of nearly isotropic distributed comets with aphelia of roughly 20 000 AU.

The cloud of cometary like bodies that encloses spherically the solar system at a distance of about 50 000 AU (about 0.75 light year) is named Oort cloud. Newer investigations have revealed that it is in fact composed of two parts:

- outer part: spherical, 20 000–50 000 AU,
- inner part: more disc-shaped, 2000–20 000 AU, also known in the literature as Hill's cloud.

The Oort cloud is the source of long period comets. Especially the objects in the outer cloud are only loosely bound to the Solar System and can be easily disturbed by passing stars.

The mass of the outer cloud is assumed to be about 5 times the mass of the Earth, the number of objects larger than about 1 km is several 10^9. The majority of the objects consist of ices such as water, methane, ethane, carbon monoxide and hydrogen cyanide. There are also indications, that rocky objects may exist there.

Origin of the Oort Cloud

To explain the origin of the Oort cloud, we must go back to the time of the formation of the solar system 4.6 billion years ago. The population of the Oort was originally

[3] Because of turbulence in the Earth's atmosphere it is extremely difficult to observe such small angular separations.

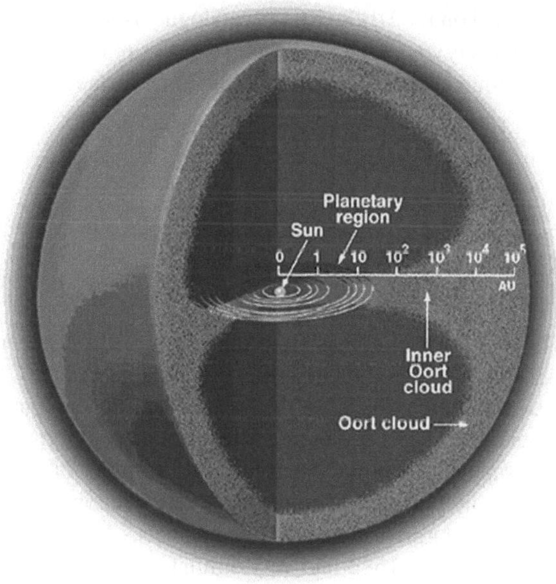

Fig. 5.2 The Oort cloud named after the Dutch astronomer J.H. Oort who proposed its existence in 1950. Planetary are orbits not in exact scale

formed in the inner solar system like the planets but due to the strong gravity of the gas giants (mainly Jupiter), these objects were ejected into long elliptic or parabolic orbits.

The formation of the Oort cloud was simulated by computer models (Dybczynski et al., 2008 [112]). The disk was divided from 4 to 50 AU into $(50-4)/\Delta a$ concentric rings. The number of particles in the outer ring with a semi-major axis equal to $(50-\Delta a/2)$ is assumed to be N_{50}, the number N_a with semi-major axis a is

$$N_a = \sqrt{\frac{50-\Delta a/2}{a}} N_{50}$$

For the simulations, the following values were taken: $\Delta a = 0.1$ AU and $N_{50} = 14$, therefore in total 10 038 objects were considered. These objects were subjected to different forces:

- four giant planets,
- galactic tides,
- nearby passing stars.

The population size of the Oort cloud reaches a maximum value at about 210 Myr and becomes stable from about 500 Myr. At 1 Gyr, the population size reaches 40% of its maximum value. Only 0.3% of the initial particles resided in the Oort cloud after that time interval.

Nihei et al., 2007 [247] reported on the detectability of occultations of stars by objects in the Kuiper Belt and Oort Cloud.

Based on the HIPPARCOS catalogue, Bobylev, 2010 [34] made a search for stars that closely encounter with the solar system in the interval −2 Myr and +2 Myr. They found nine candidates within 30 pc distance.

5.2 Comets

5.2.1 Early Observations

The word comet comes from the Greek word *kome* meaning the hair of the head. Aristotle described them as stars with hairs, *kometes*. Comets were usually considered bad omens because they seemed to appear randomly. In 1577 Tycho Brahe observed a bright comet and measured its position as did an observer at a different geographical location. Therefore, the parallax of the comet was determined. This was the first proof that comets are not a phenomenon in the Earth's atmosphere but must be at large distances like the planets.[4] In 1705 E. Halley noted that the comets of 1531, 1607, 1682 had very similar orbital parameters. Slight changes were explained by perturbation of Jupiter and Saturn. Halley was able to predict the return of the comet in 1759. Therefore, this object was named after him. The next return of Halley's comet will be in 2061.

There is no clear distinction between comets and asteroids. Very old comets have lost all their volatiles and the nucleus resembles to asteroids. There are more than 3500 reported known comets which certainly is only a tiny fraction of the whole comet population in the solar system. The main reservoir of comets is in the Oort cloud and there may be several 10^{12} cometary like objects there.

Most of the comets observed are unspectacular. Some are very bright and some are famous because they return periodically like Halley's comet.

5.2.2 Orbital Characteristics of Comets

There exist several classification schemes for comets and one criterion is to use their orbital parameters.

- Short period comets: their orbital periods about the Sun are less than 200 years. At aphelion (when their distance from the Sun is at maximum) they are at distances of Jupiter and beyond. The aphelion of comet Halley is about the orbit of Neptune. Comets of the Jupiter family have orbital periods of less than 20 years. The existence of such families of comets can be attributed to the gravitational influence of Jupiter.[5]

[4]Their measurements gave a distance four times that of the Moon.

[5]Also called Jupiter family; there exists also a Saturn family.

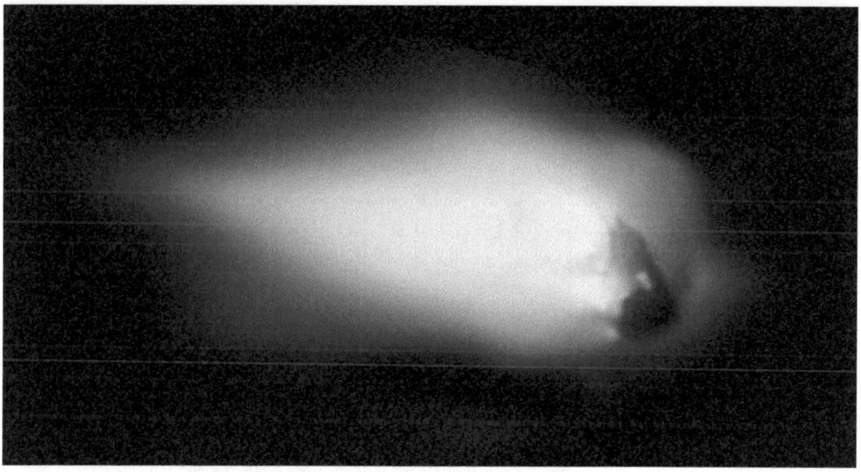

Fig. 5.3 Nucleus of comet Halley, observed from the GIOTTO mission in 1986. Courtesy: ESA, MPAe, Lindau

- Long period comets: they have eccentric orbits with periods greater than 200 years. Their orbital periods can extend several 10^3 up to 10^6 years. However, they are not on a parabolic orbit which means that they are still gravitationally bound to the Sun. Their plane of orbit may be highly inclined with respect to the ecliptic.
- Single-apparition comets: they are on a parabolic or hyperbolic orbit that means after one perihelion passage they will leave the solar system.

The icy objects in the Kuiper belt and Oort cloud become ejected into the inner solar system where they appear as comets. Perturbations by outer planets occur in the Kuiper belt, these become short period comets, perturbations by nearby passing stars affect objects in the Oort cloud, these become long period comets.

5.2.3 Physics of Comets

Comets consist of mainly three parts:

- Nucleus: composed of rock, dust, ices, frozen gases; often described as dirty snowballs, however the ices are covered by dark dust and rocks. The nuclei are irregularly shaped and the diameters range from several 100 m to some dozens of km. Up to now only the nuclei of the comets Halley (Fig. 5.3), Borrelly, Wild and Tempel 1 have been investigated by satellite missions and photographed directly.
- Coma: streams of dust and gas are released when a comet approaches the inner solar system (i.e. approximately inside the orbit of Mars) because of the increasing solar radiation. A tenuous but extended (several 10^4 km) atmosphere, called coma is formed.

Fig. 5.4 Comet Hyakutake

- Tail: When comets get closer to the Sun, a tail forms which always points in the direction opposite to the Sun because of the influence of solar wind and solar radiation pressure. There are two types of tails that point to slightly different directions. The ion tail consists of gas (plasma) points directly away from the Sun, because it is more strongly affected by the solar wind (and magnetic field since the particles are charged) than dust that is affected by the radiation pressure. The ion tail has a blueish appearance, the dust tail is brighter and curved (because of the Keplerian motion of particles). The cometary tails might become very spectacular and extend more than 1 AU (150 million km) (see Fig. 5.4).

Solar UV radiation ionizes particles in the coma. This forms an induced magnetosphere around a comet. A bow shock is formed upstream of the comet, in the flow direction of the solar wind. In this region, large concentrations of cometary ions[6] congregate and act to load the solar magnetic field with plasma; the field lines drape around the comet forming the ion tail. If the ion tail loading is sufficient, then the magnetic field lines are squeezed together to the point where, at some distance anti-sunward along the ion tail, magnetic reconnection occurs. This leads to "tail disconnection events" which were observed e.g. when the ion tail of comet Encke passed through a solar coronal mass ejection. For such events also solar coronal mass ejections (CMEs) play an important role (see e.g. Vourlidas, 2007 [348]).

Space Missions to Comets

Spacecraft missions like GIOTTO (see Fig. 5.3) (comet Halley) and Deep Space 1 (comet Borrelley) have shown that cometary nuclei are very dark (albedo less than 0.04).

[6]Pick up ions.

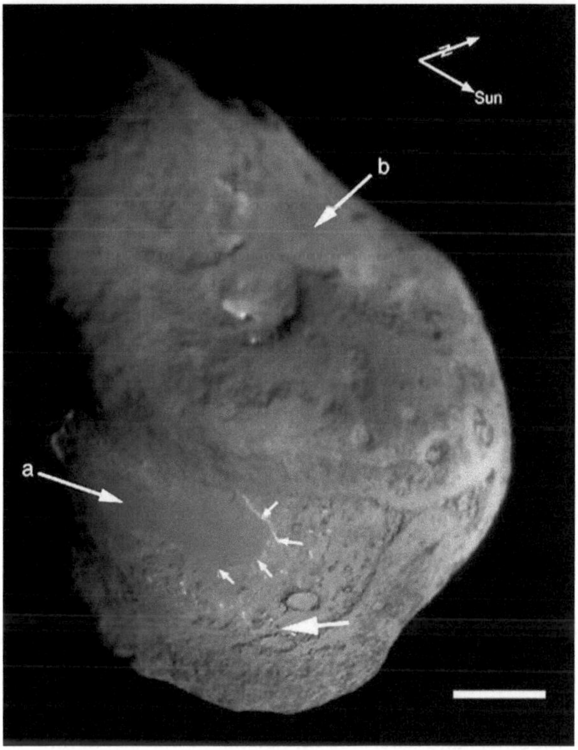

Fig. 5.5 Nucleus of comet Tempel 1 imaged by the Deep Impact impactor. The nucleus measures about 6 kilometers across. *Arrows a* and *b* point to large, smooth regions. The impact site is indicated by the third *large arrow*. Small grouped arrows highlight a scarp (a cliff or steep slope along the edge of a plateau) that is bright due to illumination angle. They show a smooth area to be elevated above the extremely rough terrain. The white scale bar in the lower right represents 1 km across the surface of the comet nucleus. The two directional arrows (vectors) in the upper right point to the Sun and Celestial North. Photo credit: NASA/UM M.F. A'Hearn et al., Science 310, 258 (2005)

The GIOTTO mission was Europe's first deep space mission. For the first time close-up images of a comet nucleus (comet Halley) were made. It was also the first spacecraft to encounter two comets and its orbit was changed by returning to Earth for a gravity assist. The launch date was on July 2, 1985. GIOTTO passed the nucleus of comet 1P/Halley at a distance of less than 600 km on March 13, 1986[7] (Fig. 5.3). A summary of the scientific results of the Giotto mission was given by Reinhard, 1988 [278].

Comet Tempel 1 was discovered in 1867 by E. Tempel. At the time of discovery, the comet was 0.7 AU from the Earth and 1.6 AU from the Sun. Currently, Tempel 1 has an orbital period of 5.5 years and a perihelion distance of roughly 1.5 AU. With an orbital eccentricity of 0.5, Tempel 1's orbit lies between the orbits of Mars and Jupiter. These values varied in the past and will change in the future because of close approaches with the planet Jupiter. It therefore belongs to the group of short period comets.

Comet Tempel 1 was selected as a target for a space mission with an impactor (see Fig. 5.5). It was also expected that due to the numerous and frequent perihelion passages, its surface and upper crust might have been considerably restructured.

[7] At that time the distance was 0.89 AU from the sun and 0.98 AU from the Earth.

The Deep Impact mission revealed many details of the properties of comet Tempel 1. The spacecraft arrived at comet Tempel 1 on July 4, 2005 to impact it with a 370 kg mass. The spacecraft was launched from Earth in January, 2004. The impactor was a battery powered spacecraft that operated only for one day. Images from the comet were obtained till few seconds before the collision. The closest approach of the main spacecraft to the comet was 500 km. The flyby spacecraft monitored the impact.

The impact also showed that the cometary nucleus is extremely porous and that the ice was close to the surface but below a devolatilized layer with thickness of order the impactor diameter. The impact crater was about 100 m the impact itself did not change the orbital parameters of the object. The surface of the nucleus revealed layered structures, signs of outbursts, many of them correlated with the rotational period and also impact craters. A review about the results was given recently by A'Hearn, 2008 [4].

5.2.4 Collisions with Comets

Collisions between comets and planets have occurred often during the evolution of the solar system. Cometary impacts on early Earth may have deposited a substantial amount of water to the Earth. The masses of lost and retained water after the impacts of comets and asteroids on oceans of various depths were studied by Svetsov, 2009 [330]. The bombardment of an atmosphereless planet by fast asteroids can wipe out the most part of an ocean. Because of their mass loss during perihelion passage, it is difficult to predict cometary orbits with high precision. Therefore an impact of a comet on a planet cannot be predicted precisely.

Due to the gravitational influences of planets, comets may even break up into several pieces. Comet Shoemaker-Levy 9 broke up into pieces and collided with Jupiter in 1994. This spectacular event was observable from Earth with even small amateur telescopes.

If the comet's path crosses Earth's path, then at that point meteor showers are observable as the Earth passes through the trail of debris. The Perseid meteor shower occurs every year between August 9 and 13 when the Earth passes through the orbit of the comet Swift-Tuttle. Halley's comet is the source of the Orionid shower in October.

Many comets were detected with the solar satellite SOHO[8] (sungrazing comets), they approach the Sun to within several 10^5 km or even closer and some of them become evaporated during this approach others may survive several passages.

The loss of the surface layers of comet nuclei was investigated by Thomas, Alexander and Keller, 2008 [334]. Their conclusion is that local variations in outgassing rates can be substantial, subsurface sublimation down to a depth of 2.3 cm.

[8] Solar and Heliospheric Observatory.

5.2.5 *Detection of Water on Comets*

As we have already pointed out, comets may have been very important for the Earth and terrestrial planets because they deposited during collisions considerable amounts of water on the surfaces of these planets. On March 18, 1988 dark features on nine consecutive photographs were observed on Venus. Since film defects and other interferences (e.g. from an artificial Earth satellite or interplanetary object) can be ruled out, it is highly probable that this event was an impact of a small cometary like object that took place on the upper haze layer of the dense Venusian atmosphere. Because such an object consists mainly of water, evaporation of H_2SO_4 particles occurred which decreased the albedo at the point of entrance and therefore a dark feature appeared (Kolovos, Varvoglis and Pylarinou, 1991 [189]).

Water has strong absorption bands around 1.45, 1.95 and 2.50 μm. Mumma et al. 1986 [241] reported on the detection of water vapor in Halley's comet. Nine spectral lines around 2.65 μm were found by means of a Fourier transform spectrometer on the NASA-Kuiper Airborne Observatory.[9] The water production rate was about 6×10^{28} molecules per second on 22.1 December and 1.7×10^{29} molecules per second on 24.1 December UT in 1985.

Ground based detection of gaseous water in the coma of comet Halley was reported by Knacke et al., 1986 [185]. They observed at 1.4 and 1.9 μm and derived a mass loss rate from the intensities observed of 2×10^{29} molecules/s. The detection of water group ions from comet Halley by means of Sakigake[10] was described by Oyama and Abe, 1990 [257]. During the impact of Shoemaker Levy on Jupiter, water was detected in the resulting fireball of the fragments G and K (Bjoraker et al., 1994 [31] and 1996 [32]).

Davies et al., 1997 [96] investigated spectra of Comet Hale-Bopp (C/1995 O1) covering the range 1.4–2.5 μm that were recorded when the comet was 7 AU from the Sun and they found broad absorption features at 1.5 and 2.05 μm. Much of the light from the comet is scattered from the coma, some of the absorption features can be matched by an intimate mixture of water ice and a low-albedo material such as carbon on the nucleus. Furthermore, the absence of the 1.65 μm absorption feature of crystalline ice suggests that the cometary ice was probably in an amorphous state at the time of these observations.

Comsovici et al., 1998 [78] observed Comet Hyakutake C/1996 B2 during the period 26 April–4 May 1996 around perihelion searching for the 22 GHz neutral water line by using a fast multichannel spectrometer coupled to the 32 m dish of the

---

[9]The Kuiper Airborne Observatory (KAO) was a national facility operated by NASA to support research in infrared astronomy. The observation platform was a highly modified C-141A jet transport aircraft capable of conducting research operations up to a height above surface of 14 km. The KAO's telescope was a conventional Cassegrain reflector with a 36-inch aperture, designed primarily for observations in the 1 to 500 μm spectral range. Its flight capability allowed it to rise above almost all of the water vapor in the earth's atmosphere allowing observations of infrared radiation, which is blocked before reaching ground-based facilities.

[10]A Japanese spacecraft that approached comet Halley to within 7 million km in 1985.

Fig. 5.6 The Medicina 32 m radiotelescope. The 32 m parabola is used both in single dish and for interferometric observations. The frequency coverage is between 1.4 and 22 GHz

Medicina radiotelescope (Fig. 5.6).[11] They detected water in this comet when it was only 0.23 AU from the Sun. Moreover, the lines observed were split and showed velocities varying from 22 to 44 km/s. Normally, the outflow velocity of neutral molecules from cometary nuclei is expected at about 1 km/s. These high values of the velocities observed may be explained by the strong solar coronal activity which could have accelerated icy grains.

Water was also detected on the comets 71P/Clark and C/2004 B1 (Linear) with the Spitzer Space Telescope (Bockelée-Morvan, 2009 [35]).

A dimer is a compound composed of two units (monomers), which may react either by

- addition,
- condensation.

Crifo, and Slanina, 1991 [85] studied the formation of water dimers in a cometary atmosphere (by condensation processes). They obtained a partial recondensation of the coma water into large clusters.

Schulz et al., 2006 [300] reported on the detection of water ice grains after the DEEP IMPACT onto Comet 9P/Tempel 1.

[11]Located 30 km from Bologna and operated by the INAF (National Institute for Astrophysics).

5.3 Asteroids

5.3.1 General Properties

On the first day of January 1801, Giuseppe Piazzi discovered an object which he first thought was a new comet. But after several observations its orbit was precisely determined, and it became clear that this object was not a comet but was in orbit around the Sun like a planet, however evidently the size of the object was smaller. Therefore it was called an asteroid and named Ceres.[12] Three other small bodies were discovered in the next few years (Pallas, Vesta, and Juno). By the end of the 19th century several hundred asteroids were known.

Several thousand asteroids have been discovered and given provisional designations so far. Thousands more are discovered every year. There are 26 known asteroids larger than 200 km in diameter. About 99% of all objects >100 km are known, however, of the total number of asteroids with diameters between 10 and 100 km we know only 50%. It is difficult to estimate the total number of asteroids, perhaps as many as a million 1 km sized asteroids may exist most of them being too small to be detectable from Earth.

Since most of the asteroids have orbits between Jupiter and Mars, it was first assumed that they are remnants of a larger planet that broke up. However, the total mass of all the asteroids is less than that of the Moon.[13]

Ceres has a diameter of 933 km, the next largest are Pallas, Vesta and Hygiea which are between 400 and 525 km in diameter. All other known asteroids are less than 340 km. Smaller asteroids are irregularly shaped as the asteroid Gaspra ($20 \times 12 \times 11$ km) (Fig. 5.7).

5.3.2 Classification of Asteroids

Asteroids are classified into:

- C-type: extremely dark (albedo 0.03), similar to carbonaceous chondrite meteorites; 75% of known asteroids belong to this class.
- S-type: 17% of asteroids; bright (albedo 0.1–0.2); metallic Ni, Fe and Mg silicates.
- M-type: bright (albedo 0.1–0.2), pure NiFe.
- Rare types.

One should however take into account biases in the observations—e.g. dark C-types are more difficult to detect.

According to their position in the solar system, asteroids can also be categorized into:

---

[12]According to the new definition of the International Astronomical Union, IAU, adopted in 2006, Ceres and Pluto are classified as dwarf planet.

[13]The mass of the moon is only $1/81$ Earth masses.

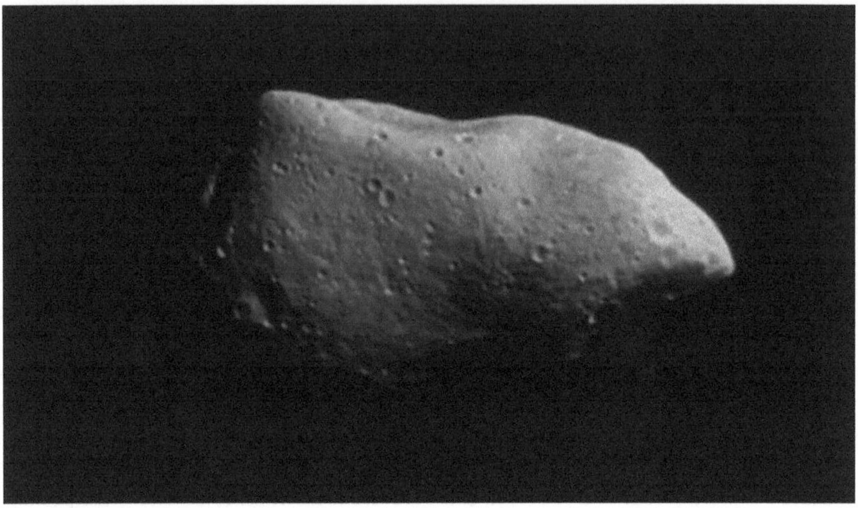

Fig. 5.7 Asteroid 460 Gaspra. Note the shape that strongly differs from spherical

- Main belt: located between Mars and Jupiter, 2–4 AU from the Sun.
- Near Earth Asteroids (NEAs): they closely approach the Earth and will be treated separately.
- Trojans: located near Jupiter's Lagrange points (60 degrees ahead and behind Jupiter in its orbit); several hundreds are known.
- Between the main concentration in the Main Belt are relatively empty regions known as Kirkwood gaps. These are regions were an object's orbital period is a simple fraction of that of Jupiter (resonance).
- Centaurs: asteroids in the outer solar system; e.g. Chiron (his orbit lies between Saturn and Uranus).

The giant outer planets formed from an agglomeration of billions of comets. The comets we observe today are the left over pieces like the asteroids are the left over pieces from the formation of the inner planets (Mercury, Venus, Earth and Mars).

5.3.3 NEOs

Near-Earth Objects (NEOs) are comets and asteroids that have been nudged by the gravitational attraction of nearby planets into orbits that allow them to enter the Earth's neighborhood. Composed mostly of water ice with embedded dust particles, comets originally formed in the cold outer planetary system while most of the rocky asteroids formed in the warmer inner solar system between the orbits of Mars and Jupiter.

In terms of orbital elements, NEOs are asteroids and comets with perihelion distance q less than 1.3 AU. Near-Earth Comets (NECs) are further restricted to include

Table 5.1 Groups of Asteroids near Earth orbit. MOID means mean orbit intersection distance from Earth, PHAs are potentially hazardous asteroids

Group	Description	Definition
NECs	Near-Earth Comets	$q < 1.3$ AU, $P < 200$ years
NEAs	Near-Earth Asteroids	$q < 1.3$ AU
Atens	Earth-crossing NEAs	$a < 1.0$ AU, $Q > 0.983$ AU
Apollos	Earth-crossing NEAs	$a > 1.0$ AU, $q < 1.017$ AU
Amors	Earth-appr. NEAs with orbits betw. Earth and Mars	$a > 1.0$ AU, $1.017 < q < 1.3$ AU
PHAs	NEAs	MOID ≤ 0.05 AU, H ≤ 22.0

only short-period comets (i.e. orbital period P less than 200 years). The vast majority of NEOs are asteroids, referred to as Near-Earth Asteroids (NEAs). NEAs are divided into groups (Aten, Apollo, Amor) according to their perihelion distance (q), aphelion distance (Q) and their semi-major axes (a) as is listed in Table 5.1.

5.3.4 The Cretaceous-Tertiary Impact

At the Cretaceous-Tertiary boundary (K/T) about 65 million years ago, an impact occurred. Mass extinction of a broad spectrum of lifeforms (Raup and Sepkoski, 1988 [275]), a worldwide clay layer containing geochemical (Alvarez et al., 1980 [8]), mineralogical (Bohor, 1990 [36]) and isotopic anomalies (McDougall, 1988 [224]) and tsunami deposits (Bourgeois et al., 1988 [41]) point to such a catastrophic event at that time.

The buried Chicxulub basin is an ancient 300 km diameter impact crater buried underneath the Yucatán Peninsula in Mexico. There is evidence that the Chicxulub crater has been formed by a long-period comet composed primarily of nonsilicate materials (ice, hydrocarbons etc.) and subordinate amounts of primitive chondritic material. The collision would have raised the energy equivalent to between 4×10^8 and 4×10^9 megatons of TNT. Studies of terrestrial impact rates suggest that such an event would have a mean occurrence rate of $\sim 1.25 \times 10^{-9}$ yr^{-1}. This rate is considerably lower than that of the major mass extinctions over the last 250 million years ($\sim 5 \times 10^{-7}$ yr^{-1}). The crater showed several rings (similar to the rings of the Mare Orientale on the Moon).

In Fig. 5.8 the estimated frequency of impacts as a function of asteroid diameter is shown. An asteroid or comet ≥ 10 km in diameter (which releases $\geq 10^{24}$ J or 10^8 Mt TNT) would cause a global catastrophe. Also the K/T event is pointed out in the figure.

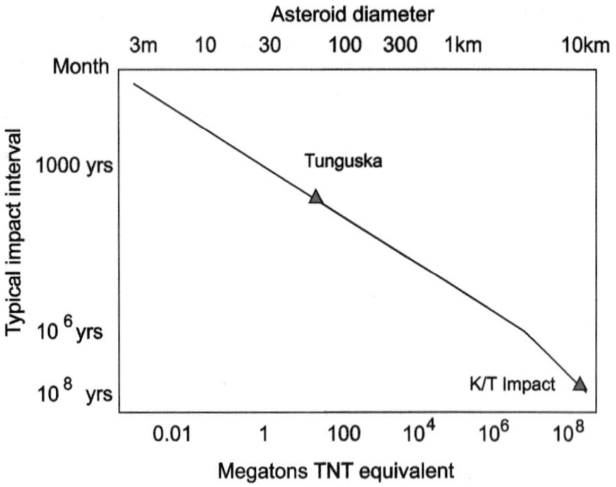

Fig. 5.8 Estimated frequency of impacts on the Earth from the present population of comets and asteroids and impact craters

5.3.5 Water and Ice on Asteroids

The object 1 Ceres that is now classified as dwarf planet was the first asteroid where spectral signatures of water were found. Near 3.0 microns an absorption feature indicates the presence of water. From that it was estimated that between 10 and 15% water of hydration may be found on Ceres' surface (Lebofsky, 1978 [195]). This was the first evidence of water substance on the surface of an object in the main asteroid belt between Mars and Jupiter. Hydration means that a water molecule is added to the crystal structure of a mineral. This creates a new mineral, usually called a hydrate.[14]

Lebofsky et al., 1981 [196] pursued these studies of Ceres by high-resolution Fourier spectra (1.7–3.5 microns) and medium-resolution spectrophotometry (2.7–4.2 microns). The 3-micron feature is compared with the 3-micron bands due to water of hydration in clays and salts and they conclude that the spectrum of Ceres shows a strong absorption at 2.7–2.8 microns due to structural OH groups in clay minerals. The dominant minerals on the surface of Ceres are therefore hydrated clay minerals structurally similar to terrestrial montmorillonites. There is also a narrow absorption feature at 3.1 microns which is attributable to a very small amount of water ice on Ceres.

Fanale and Salvail, 1989 [126] studied the presence of water ice on Ceres based on models assuming temperatures at the surface and subsurface, water fluxes, ice depths, water supply rates as a function of time, latitude and regolith properties. It was found that ice could persist in a transient state only above 80 degrees latitude.

A study of the spectral signature of water described above showed that 66 percent of the C-class asteroids in the sample investigated have hydrated silicate surfaces. Differences in hydration among different types of asteroids may be understood by the fact that the early Sun had a very strong wind which heated the protoasteroids.

[14]E.g. by the process of hydration Portland cement develops strength.

This effect was less pronounced for the more distant particles (Jones et al., 1990 [175]). Ices of volatile elements should be more abundant in the external region of the belt, where pristine materials have not undergone drastic modification processes after the accumulation in the planetesimal swarms between those of Mars and of Jupiter. There is no evidence of any water, water ice or aqueous alteration materials on the D-type asteroids, which are considered the least altered objects (Barucci, Fulchignoni and Lazzarin, 1996 [19]). D-type asteroids have a very low albedo and appear reddish and are found in the outer asteroid belt. It has been suggested that they have a composition of

- at the surface: organic rich silicates, carbon and anhydrous silicates,
- in the interior: possibly with water ice.

Yang and Jewitt, 2007 [368] searched for spectral signatures of water ice on Trojan asteroids.[15] Trojans (911) Agamemnon, (617) Patroclus, (1143) Odysseus, and (2797) Teucer were observed. These objects have been independently reported to show possible weak absorptions at 1.7 and 2.3 μm, respectively. If real, the latter features may be due to organic materials present on the surfaces. Trojan (4709) Ennomos has a geometric albedo significantly above the mean Trojan albedo, perhaps because a recent impact has coated part of the surface with freshly excavated ice.

Eucrites are achondritic stony meteorites that originate from the surface of the asteroid 4 Vesta. The meteorite Serra de Magé, an eucrite, contains quartz veinlets. They are identical to 'crack-seal' quartz veinlets in terrestrial rocks, and are extraterrestrial and ancient because they pre-date a 4.40 Ga metamorphism. The quartz was likely deposited from liquid water solutions (as are terrestrial veins). Because there is no indication of internal (magmatic) water in the eucrite meteorites and thus in Vesta, the water from which the veinlet was deposited probably came from outside Vesta. By analogy with water ice deposits on the Moon and Mercury, Vesta and similar asteroids may have had (or now have) polar ice deposits, possibly remainders from comet impacts (Treiman et al., 2004 [339]).

Water ice on the surface of asteroid 24 Themis was detected by IR observations (Campins et al., 2009 [55]). The problem is that water ice cannot be stable on the surface of Themis at 3.2 AU over the last 4.5 Gy. Maybe the solution is that of an activated asteroid.

Lightcurves (rotation) can be due to albedo effects (e.g. Pluto) or shape effects (asteroids). With the IR astronomical satellite Herschel Trans Neptunion objects, TNOs, can be observed and from their lightcurves in the infrared, e.g. at 100 and 160 μm and the comparison of these lightcurves with their known optical lightcurves information about the thermal properties on their surfaces can be obtained (Mueller et al., 2010). A correlation of the thermal lightcurve with the visible light curve indicates a shape effect (the body is irregular, therefore the light variations).

---

[15] This group of asteroids are found near the equilibriums points L4 and L5 in the Sun-Jupiter system.

5.3.6 Asteroids as a Source for Water on Earth

Numerical studies were made in order to investigate the evolution of terrestrial planets and asteroids in the early solar system. Here we give some results of Petit et al., 2000 [260].

In the early solar system, planetary embryos spread between 0.5 and 4 AU were assumed. At that time Jupiter was already formed. There occurred numerous mutual interactions between these planetesimals and also with the still growing Jupiter and collisions happened frequently. Planets were formed in the terrestrial planet regions and within 1 million years 99% of the embryos were ejected by the gravitational influence of Jupiter. Only 1% of the original population survived in the 2.8–3.3 AU region which is now known as the main asteroid belt. The surviving particles have undergone changes in semi-major axis of several tenth of an AU, which could explain the radial mixing of asteroid taxonomic types. Some of the particles ended up on very inclined eccentric orbits in the inner Solar System, on orbits with a longer decay time. These particles could be the source of the Late Heavy Bombardment. In this scenario, the Earth would continuously accrete water during its formation, from the earliest phases when the solar nebula was still present, to the late stages of gas-free sweepup of scattered planetesimals. Asteroids and the comets from the Jupiter-Saturn region were the first water deliverers, when the Earth was less than half its present mass. The bulk of the water presently on Earth was carried by a few planetary embryos, originally formed in the outer asteroid belt and accreted by the Earth at the late stages of its formation. Finally, a late veneer, accounting for at most 10% of the present water mass, occurred due to comets from the Uranus-Neptune region and from the Kuiper belt.

The deuterium to protium (D/H) ratio of the deep mantle may be a remnant of the hydrogen isotopic composition of Earth forming planetesimals. The mass of asteroids and comets incident on Earth since the time of its accretion is estimated to be 4×10^{20}–2×10^{22} kg. The combined use of water D/H ratios, the lunar cratering record, and terrestrial mantle siderophiles would favor a rather low mass fraction of comets among impacting bodies (≤ 0.01). Asteroids, comets, and the early Earth contributed to 0–0.5, 0–0.1, and 0.5–0.9 of Earth's water inventory, respectively (Dauphas, Robert, and Marty, 2000 [94]).

The role of comets and primitive asteroids in supplying water and prebiotic organics to early Earth was reviewed by Mumma, 2008 [240].

5.4 Meteorites

5.4.1 General Properties

The term meteor comes from the Greek *meteoron*, meaning phenomenon in the sky. Meteors are small solid particles that enter the Earth's atmosphere from interplanetary space. They move at high speeds and the friction they encounter in the air vaporizes them (typically at heights between 80 and 110 km above the surface). The light caused by the luminous vapors formed in such an encounter appears like a star moving rapidly across the sky, fading within a few seconds. To be visible, a meteor must be within 200 km of the observer. The total number of meteors bright enough to be visible is estimated to be about 25 million per day. A typical bright meteor is produced by a particle with a mass less than 1 g. A particle the size of a golf ball produces a bright fireball. The total mass of meteoritic material entering the Earth's atmosphere is estimated to be about 100 tons per day.

A meteoroid is matter revolving around the sun or any object in interplanetary space that is too small to be called an asteroid or a comet. Even smaller particles are called micrometeoroids or cosmic dust grains, which includes any interstellar material that should happen to enter our solar system. A meteorite is a meteoroid that reaches the surface of the Earth without being completely vaporized.

One of the primary goals of studying meteorites is to determine the history and origin of their parent bodies.

On Earth, meteorites from the Moon and Mars may be found. Several achondrites sampled from Antarctica since 1981 have conclusively been shown to have originated from the moon. This is proven by compositional matches of lunar rocks obtained by the Apollo missions of 1969–1972 with these achrondrites. Sources of other specific meteorites remain unproven, although another set of eight achondrites are suspected to have come from Mars. These meteorites contain atmospheric gases trapped in shock melted minerals which match the composition of the Martian atmosphere as measured by the Viking landers in 1976. All other groups are presumed to have originated on asteroids or comets; the majority of meteorites are believed to be fragments of asteroids.

5.4.2 Classification

Meteorites can be classified into stony, stony iron and iron. The most common meteorites are chondrites which are stony. Radiometric dating indicate an age of about 4.5×10^9 years. Achondrites are also stony but they are considered differentiated or reprocessed matter. They are formed by melting and recrystallization on or within meteorite parent bodies. Pallasites are stony iron meteorites composed of olivine enclosed in metal.

The motion of meteoroids can be severely perturbed by the gravitational fields of major planets. Jupiter's gravitational influence is capable of reshaping an asteroid's orbit from the main belt so that it dives into the inner solar system and crosses the orbit of Earth. This is apparently the case of the Apollo and Vesta asteroid fragments.

Particles found in highly correlated orbits are called stream components and those found in random orbits are called sporadic components. It is thought that most meteor streams are formed by the decay of a comet nucleus and consequently are spread around the original orbit of the comet. When Earth's orbit intersects a meteor stream, the meteor rate is increased and a meteor shower results. Typically, a meteor shower will be active for several days. A particularly intense meteor shower is called a meteor storm. Sporadic meteors are believed to have had a gradual loss of orbital coherence with a meteor shower due to collisions and radiative effects, further enhanced by gravitational influences. There is still some debate concerning sporadic meteors and their relationship with showers. A well known meteor shower are the Perseids (named after the radiant that is the name of the constellation where the meteorites are seem to be coming from) which has its maximum on August 11.

5.4.3 Water in Meteorites

Let us briefly list some minerals that are frequently found in meteorites:

- Olivine: a magnesium iron silicate.The formula is:

$$(Mg, Fe)_2SiO_4 \tag{5.2}$$

  Olivine is one of the most common minerals on Earth, also found in meteorites, on the Moon, Mars, etc. and it is of olive green color. Transparent olivine is sometimes used as a gemstone called peridot.
- Pyroxene: group of important rock-forming silicate minerals found in many igneous and metamorphic rocks; presence in volcanic lavas, where they are sometimes seen as crystals embedded in volcanic glass.

Ashworth, and Hutchison, 1975 [11] made electron microscopic observations of the hydrous alteration products of olivine in an achondrite and in an ordinary chondrite. Their conclusion was that the Nakhla achondrite, and possibly the Weston chondrite, contain water of extraterrestrial origin which was mobilized by mild shock deformation. Carbonaceous chondrites are believed to be unaltered material left over from the formation of the solar system. They contain substantial amounts of reduced carbon and of water in the form of hydroxyl ions. The oxidation state of iron in some carbonaceous chondrites has been determined by means of Moessbauer spectroscopy, and it is demonstrated that there is a correlation between the oxidation state of iron and the content of water and reduced carbon in the meteorites (Roy-Poulsen et al., 1981 [284]).

The Shergotty-Nakhla-Chassigny (SNC) meteorites are believed to be of Martian origin. They contain 0.04–0.4% water by weight. The question is whether this water is of terrestrial or extraterrestrial origin (Karlsson et al., 1991 [179]). Oxygen isotope measurements can be used to answer this question. A detailed analysis of water in SNC meteorites showed:

- A portion of water was not in oxygen isotopic equilibrium.
- This implies that the lithosphere and the hydrosphere of the parent body formed two distinct oxygen isotopic reservoirs.

The conclusion of Karlsson et al., 1992 [180] was that if Mars was the parent body, the maintenance of two distinct reservoirs may result from the absence of plate tectonics on the planet.

Water contents and H isotopic compositions have been determined with the CRPG-Nancy ion microprobe in twelve chondrules (type I and type II after McSween's classification) from several deuterium-rich LL3 meteorites. The water concentration ranges from 500 to 17 000 ppm, with most of the data lying between 1000 and 5000 ppm (Deloule and Robert, 1996 [103]).

Ion probe determinations of water concentrations and D/H ratios have been performed on different phases (pyroxene, olivine, and mesostasis) of chondrules from the Bishunpur and Semarkona LL3 chondrites. Mean water concentrations average 1290 ppm in olivine (500 to 2100 ppm), 2400 ppm in pyroxene (400 to 9800 ppm), and up to 3950 ppm in the mesostasis (1000 to 16 500 ppm) (Deloule, Robert and Doukhan, 1998 [104]). The D-rich component is interpreted as a preserved interstellar source, while the low D/H ratio reflects a water component that underwent isotopic exchange with protosolar molecular hydrogen.

Chapter 6
Water on Extrasolar Planets?

Up to 2010 about 450 extrasolar planets have been detected including cases of systems with more than one planet. Due to limited observational capabilities at present, theses observations are strongly biased, as it will be discussed when we will shortly describe some methods of detection of extrasolar planets.

6.1 How to Detect Extrasolar Planets

Since planets only reflect light from their central star they are extremely difficult to detect:

- The radiative flux from the central star is orders of magnitudes higher than the radiation of the orbiting planet.
- As seen from Earth, an extrasolar planet is generally quite close to the central star, its angular separation is extremely small.
- These observational difficulties cause strong bias since it is more likely that massive planets will be detected than small faint ones.

The fact that 1995 marked the discovery of the first extrasolar planet can be compared with the Copernican Revolution that not the Earth but the Sun is the center of our planetary system and the discovery of Hubble, that the universe is expanding. Since 1995 it became evident that planetary systems are common in the universe and this has certainly big impact on the question whether there are other habitable planets or not.

6.1.1 Detection Methods

The following methods are applied in order to detect extrasolar planets:

A. Hanslmeier, *Water in the Universe,* Astrophysics and Space Science Library 368,
DOI 10.1007/978-90-481-9984-6_6, © Springer Science+Business Media B.V. 2011

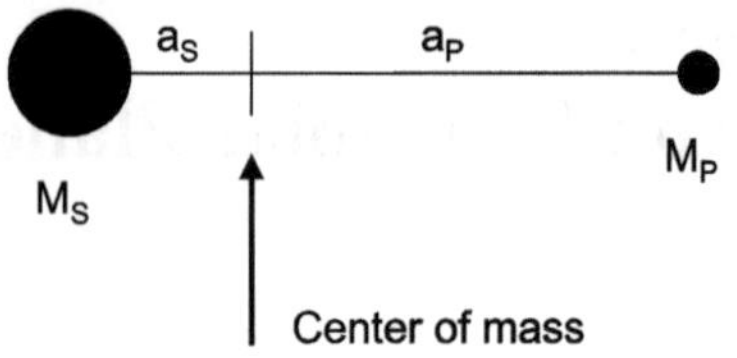

Fig. 6.1 Center of mass in a system of a star with mass M_S and a planet with mass M_P

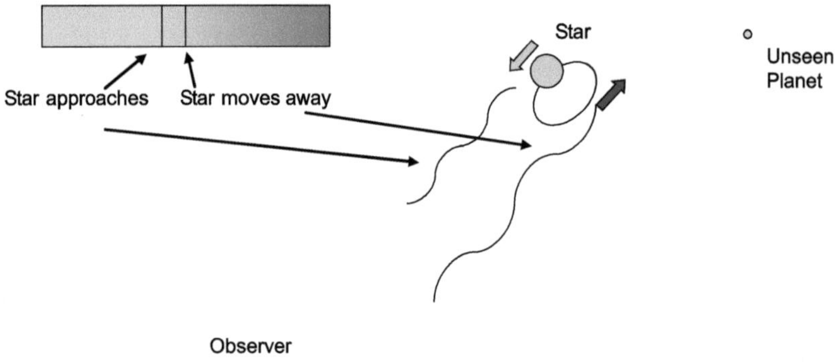

Fig. 6.2 Principle of radial velocity measurement. Note that the spectral lines of the star are blueshifted when it approaches towards the observer and redshifted when its distance to the observer increase

- Astrometry: consider two bodies with masses m_1, m_2. Moving around their center of mass (barycenter), they will attract each other according to Newton's law of gravity:

$$F = G\frac{m_1 m_2}{r^2} \tag{6.1}$$

  The more massive one of the two components, the more the center of mass will be located near to that mass (see Fig. 6.1). Let us consider the center of mass of the system Sun and Jupiter. Since the mass of the Sun is about 1000 times the mass of Jupiter, the center of mass of such a two body system must be located 1000 times nearer to the Sun than to Jupiter—in fact it is located just outside the Sun's sphere. Therefore, precise position measurements are required in order to detect the small motion of a star around the center of gravity in case of the presence of other planets.
- Radial velocity or Doppler method: as we have seen above, a star will move under the gravitational attraction of a planet. This motion about the center of gravity of the system can be detected by small Doppler shifts of the star's spectral lines (see Fig. 6.2). Note that only the radial velocity component, that is the velocity that is directed toward (or away) from the observer causes a Doppler shift which varies periodically. This method has been so far the most promising one and most of the extrasolar planets detected so far were found from these Doppler shifts. The inclination i of the orbital plane with the sky is unknown, therefore, we measure the velocity

$$v_r = v \sin i \tag{6.2}$$

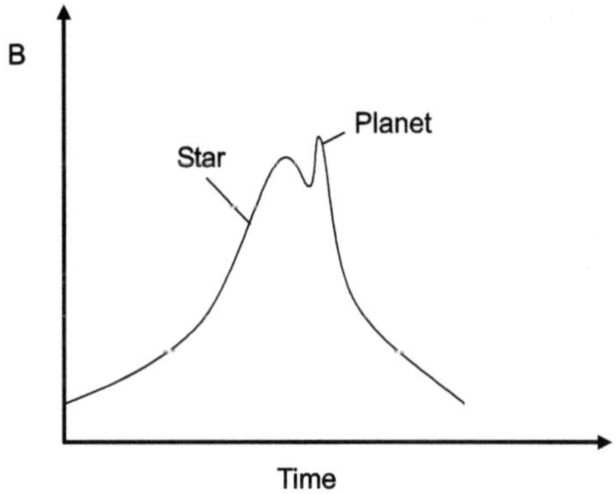

Fig. 6.3 Variation of brightness B due to the microlensing effect of a star and its planet

where v denotes the true velocity and v_r the radial velocity. When i is equal to 90° we see the orbit edge on. It is evident that small angles i lead also to low values for the measured radial velocities implying small Doppler shifts. This reduces dramatically the detection probability of planets with small i. A further complication is that one has to wait quite a long time to find periodic tiny Doppler shifts. In the case of the system Sun-Jupiter the period is 12 years and the Sun's radial velocity around the center of the Sun-Jupiter system is 14 m/s.

When we want to detect planets of Earth masses then the velocity variations would become about 1/300 that of Jupiter since the Earth's mass is only 1/300 M_J (M_J denotes the mass of Jupiter). It is therefore necessary to measure velocities in the range of cm/s. This seems unrealistic but keep in mind that for the Sun such velocity measurements are already done at regular basis.

- Transit method. planetary transits occur only when a planet crosses in front of its parent star's disk seen from Earth. If the planet is big enough then a small drop in the brightness of a star can be observed. The amount of the decrease in brightness depends on the size of the planet and the size of the star as well as on the distance of the planet from the star. Therefore, it is more probable to observe transits of giant planets that are relatively close to their central star.
- Gravitational microlensing: microlensing occurs when the gravitational field of a star acts like a lens, magnifying the light of a distant background star. If the foreground lensing star has a planet, then that planet's own gravitational field can make a detectable contribution to the lensing effect. This can be observed in the variation of the lightcurve as a secondary bump (see Fig. 6.3).
- Light reflected by a planet: it is clear that a planet orbiting a star scatters back into space some amount of the starlight it receives.

On March 6, 2009 the Kepler mission was launched (Fig. 6.4). It will sample stars in the solar neighborhood in the Orionarm of our spiral Galaxy. Transits will be observed. Such transits of exoplanets last typically between 2 and 6 hours, the amplitude of brightness variation is on the order of several 100 ppm.[1] The diameter of the telescope used is 0.95 m. A large field of view on the sky will be observed

[1] Parts per million.

Fig. 6.4 The Kepler telescope which is mainly used as a photometer to observe simultaneously the brightness of 100 000 stars using 42 CCD cameras. Credit: NASA, Kepler mission

(more than 100 square degrees, this corresponds to the area of your hand held at arm's length). In such a field the brightness variations of about 100 000 stars will be observed. Besides finding new exoplanets, also the mircrovariability of stars can be studied (e.g. Basri et al. 2008 [21]). The resolution of the dataset will be 30 minutes in time and 20 micromag in brightness.

An example of a direct (possible) observation of a planet around a brown dwarf is given in Fig. 6.5.

6.1.2 Extrasolar Planets Found by Different Detection Methods

As of Mid of June 2010 in total 461 planets were found:

1. Candidates detected by radial velocity or astrometry:
   - 369 planetary systems

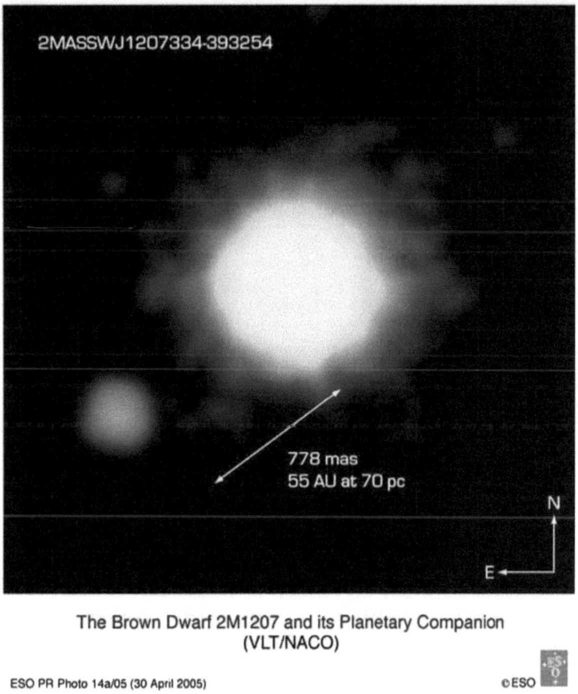

Fig. 6.5 This composite image shows the first planet outside of our solar system (*left*) found orbiting a brown dwarf, dubbed 2M1207 (*centre*). VLT/NACO. Credit: ESO

- 430 planets
- 41 multi planet systems.

2. Candidates detected by transit:

   - 87 planetary systems
   - 87 planets
   - 4 multiple planet systems.

3. Candidates detected by microlensing: 9 planets.
4. Candidates detected by imaging: 11 planetary systems, 13 planets, 1 multiple system.
5. Candidates detected by timing:

   - 5 planetary systems
   - 8 planets
   - 2 multiple planetary systems.

These data were extracted from the interactive data catalog of extrasolar planets (maintained by Jean Schneider (CNRS-LUTH, Paris Observatory)). There are also unconfirmed observations and so called cluster or free floating candidates for planets. Candidates for free floating planets are S Ori 68 (Mass 5 M_J, distance 440 pc), S Ori 70 (Mass 3 M_J, temperature 1000 K, Radius 1.6 R_J, distance 440 pc) and Cha 110913 (Mass 8 M_J, T 1300 K, radius 1.8 R_J distance 50 pc. Further candidates are CAHA Tau1, CAHA Tau2, ρ Oph 4450.

Table 6.1 Some examples of extrasolar planets. For comparison the values of Jupiter is also given. M_{ast} is the mass of the host star (in solar masses), P_{Orb} the orbital period, a the semi major axis (in AU), e the eccentricity, M_P the planet's mass and Ω the planet's rotation period

Name	Environment	P_{Orb}	a	e	M_P	Ω
Planet	M_* ($M_\odot$)	(days)	(AU)		(M_{Jup})	(Hz)
Jupiter	1.0	4332.6	5.2	0.049	1.0	1.8×10^{-4}
Rho CrB b	0.95	39.65	0.23	0.028	>1.1	1.8×10^{-6}
HD 195919 b	1.02	18.3	0.14	0.05	>3.43	4.0×10^{-6}
Gl 86 b	0.79	15.78	0.11	0.046	>4.0	4.6×10^{-6}
55 Cnc b	0.95	14.65	0.11	0.02	>0.84	5.0×10^{-6}
HD 130322 b	0.79	10.72	0.088	0.048	>1.080	6.8×10^{-6}
HD 168746 b	0.92	6.41	0.066	0.0	>0.24	1.1×10^{-5}
HD 49674 b	1.00	4.94	0.06	0.00	>0.12	1.1×10^{-5}
Ups And b	1.3	4.61	0.058	0.034	>0.71	1.6×10^{-5}
51 Peg b	0.95	4.23	0.051	0.013	>0.44	1.6×10^{-5}
HD 209458 b	1.05	3.52	0.045	0.00	>0.69	2.1×10^{-5}
HD 75289 b	1.05	3.51	0.046	0.054	>0.42	2.1×10^{-5}
BD −10316 b	1.10	3.49	0.046	0.00	>0.48	2.1×10^{-5}
Tau Boo b	1.30	3.31	0.05	0.00	>4.09	2.2×10^{-5}
HD 179949 b	1.24	3.09	0.045	0.05	>0.84	2.4×10^{-5}
HD 187123 b	1.06	3.09	0.042	0.03	>0.52	2.4×10^{-5}
HD 46375 b	1.00	3.02	0.041	0.00	>0.249	2.4×10^{-5}
HD 83443 b	0.79	2.98	0.038	0.080	>0.3	2.4×10^{-5}

6.1.3 Some Examples of Extrasolar Planets

In Table 6.1 some examples of exoplanets are given. This table was adapted from Menou et al., 2003 [229]. For comparison, the values for Jupiter are listed in the first row. M_* is the mass of the central star, P_{Orb} the orbital period, a the semi major axis, e the eccentricity of the orbit, M_P the mass of the planet and Ω the rotation period.

6.2 Habitable Zones

What makes a planet habitable? In this section we will define habitability and see that there exist several zones where a planet or satellite of a giant planet may be located to provide habitable conditions for evolution of life (life in the sense we know from Earth).

6.2.1 *Habitability*

Habitability on an astronomical object can be defined as: the conditions are such that life—as we know it from Earth—could have developed. Of course this is a restriction but it seems to be that life (see Chap. 3) is restricted to (i) presence of liquid water, (ii) based on long chains of carbon. From a qualitative point of view there exist several classes of habitats (Lammer et al., 2009 [193]):

- Class I habitats represent bodies on which stellar and geophysical conditions allow Earth-analog planets to evolve so that complex multi-cellular life forms may originate.
- Class II habitats includes bodies on which life may evolve but due to stellar and geophysical conditions that are different from the class I habitats, the planets rather evolve toward Venus- or Mars-type worlds where complex life-forms may not develop.
- Class III habitats are planetary bodies where subsurface water oceans exist which interact directly with a silicate-rich core.
- Class IV habitats have liquid water layers between two ice layers, or liquids above ice.

All these classes have in common that habitability is based on the presence of liquid water. On Earth life is based on a carbon cycle which is maintained by life but also strongly influenced by increasing solar luminosity and plate tectonics. Bounamam, von Bloh, and, Franck, 2007 [40] claimed that the number of planets predicted to bear complex life to be approximately 2 orders of magnitude lower than the number predicted for primitive life forms.Their model predicted a maximum abundance of such planets around 1.8 Ga ago and allowed us to calculate the average distance between potentially habitable planets in the Milky Way.

6.2.2 *Circumstellar Habitable Zones*

Let us consider stars of different types and ask the question whether there exists a habitable zone around them or not. In this context often the term *circumstellar habitable zone* or ecosphere is used. Obviously, the Earth is located in the center of the HZ of the solar system. Let d denote the mean radius of the habitable zone in AU, L_* the bolometric luminosity[2] of a star and $\mathrm{L}_\odot$ the bolometric luminosity of the Sun, then:

$$d = \sqrt{\frac{L_*}{\mathrm{L}_\odot}} \tag{6.3}$$

so that for $L_* = \mathrm{L}_\odot$ we obtain $d = 1$. The center of the habitable zone of a star with 0.25 $\mathrm{L}_\odot$ would be at a distance of 0.5 AU from the star. A star with 2 $\mathrm{L}_\odot$ would

[2]Bolometric means a star's output over all wavelengths.

Table 6.2 Habitable zones and some stellar parameters

Spectral class	T_{eff} (K)	Life (y)	Abundance %	HZ (AU)
O6V	41 0000	10^6	4×10^{-5}	450–900
B5V	15 400	8×10^7	0.1	20–40
A5V	8200	10^9	0.7	2.6–5.2
F5V	6400	4×10^9	4	1.3–2.5
G5V	5800	2×10^{10}	9	0.7–1.4
K5V	4400	7×10^{10}	14	0.3–0.5
M5V	3200	3×10^{11}	72	0.07–0.15

have the center of the habitable zone at a distance of 1.41 AU. In Table 6.2 the habitable zones are given for stars of different spectral type (first column), different temperatures, lifetime and abundance. We see that stars with high temperatures have short main sequence lifetimes so that it is not probable that complex life might have evolved, moreover they strongly radiate in the UV. For a G5 type star like our Sun the expected lifetime is about 10 Gyr and the circumstellar habitable zone is between 0.7 and 1.4 AU. For further details see Hanslmeier, 2009 [156].

Habitable zones around F stars are larger and further away from the star than for the Sun, habitable zones around K and M stars are smaller and nearer to the star (see also e.g. Jones, Sleep, and Underwood, 2006 [173]).

6.2.3 Galactic Habitable Zone

In the early cosmos there were no elements for the formation of planets. The Sun is a star of second or even higher generation that means it formed from an interstellar cloud that was enriched by material that was processed inside of stars and released when they exploded as a supernova. In a galaxy, there is a gradient of the metal content: near the center the metal content becomes higher than far out in the spiral arms. A planetary system must be therefore close enough to the galactic center so that a sufficiently high level of metals exist. Only under these conditions rocky planets could have formed. We have discussed the necessity of heavier elements such as carbon in order to form complex molecules for life.

If the planetary system is not far enough from the galactic center, then close encounters with passing stars become more likely. This will trigger hazards from comets that move into the inner planetary systems. Moreover, outbursts from supernovae and from the supermassive black hole at the center could cause strongly enhanced shortwavelength radiation bursts destroying the complex molecules needed for life.

Observations of extrasolar planetary systems have shown also that when the metallicity of the star becomes too high more massive planets orbiting close to the stars are likely. These could have destroyed Earth-sized objects there.

In the early stages of galaxy formation the heavy elements to form terrestrial planets were only present near the center of the galaxy because there the concentration of stars was largest and some stars already had evolved and became supernovae that enriched the interstellar medium with heavier elements there. This was not a safe environment because the danger of nearby supernovae explosions was high.

Gradually, the heavy elements spread through the galaxy and terrestrial planets formed at greater and safer distances from the galactic center. The galactic habitable zone is an annular region 7–9 kpc from the galactic center. It widens with time and is composed of stars formed between 8 and 4 billion years ago. About 75% of the stars in the galactic habitable zone are older than the Sun. and therefore, assuming that the evolution of life was on similar time scales as on Earth, we belong to the "young generation" of galactic civilizations. For further details the reader is referred to the paper of Lineweaver, Fenner, Yeshe, and Brad, 2004 [203].

6.2.4 Habitable Zone Around Giant Planets

If a habitable zone is restricted to the presence of liquid water, then we must also include the interiors of giant planets and the ice covered Galilean Moons of Jupiter, or the icy satellites of Saturn and maybe even other places. Therefore, the search for life must not be restricted to the study of circumstellar habitable zones. But it seems that life requires much more than just liquid water.

Generally, for life to occur there is also a need for an energy source. In the case of circumstellar habitable zones this is provided by the host star of the system. In the cases of satellites of planets, this can be provided by other mechanisms such as tidal heating. Therefore a liquid ocean is expected for at least Europa and other satellites. Habitable moons around extrasolar planetary systems were discussed by Williams, Kasting, and, Wade, 1997 [359]. Planets of the stars 16 Cygni B and 47 Ursae Majoris were studied in detail for hypothetical satellites. Such a moon would, however, need to be large enough (>0.12 Earth masses) to retain a substantial and long-lived atmosphere, and would also need to possess a strong magnetic field in order to prevent its atmosphere from being sputtered away by the constant bombardment of energetic ions from the planet's magnetosphere.

6.3 Dust Debris Around Stars

Before the first extrasolar planets were detected, the observational evidence of dust around stars was a first hint to the existence of extrasolar planetary systems.

6.3.1 Signatures of Dust Around Stars

Disks containing dust debris are found around stars (see e.g. Fig. 6.6). These disks are believed to be equivalent to our Kuiper Belt. There seem to be two different types of such disks:

- with a broad belt, wider than about 50 astronomical units,
- narrow ones with a width of between 20 and 30 AU and a sharp outer boundary, seemingly like our own Kuiper Belt.

Stern and Shull, 1991 [325] reported on water Masers, red giants, and Oort Clouds around other stars. Solar like stars will evolve to Red Giants as soon as their hydrogen supply is exhausted in their central cores. This so-called post main sequence phase implies an increase in luminosity by a factor of 1000. Such an increase will certainly affect the Oort cloud bodies. If we assume that such clouds are common around solar-like stars, then the question arises whether we would be able to detect such processes in giant star outflows.

Possible signatures could be:

- Ubiquitous presence of H_2O and OH around post main sequence stars,

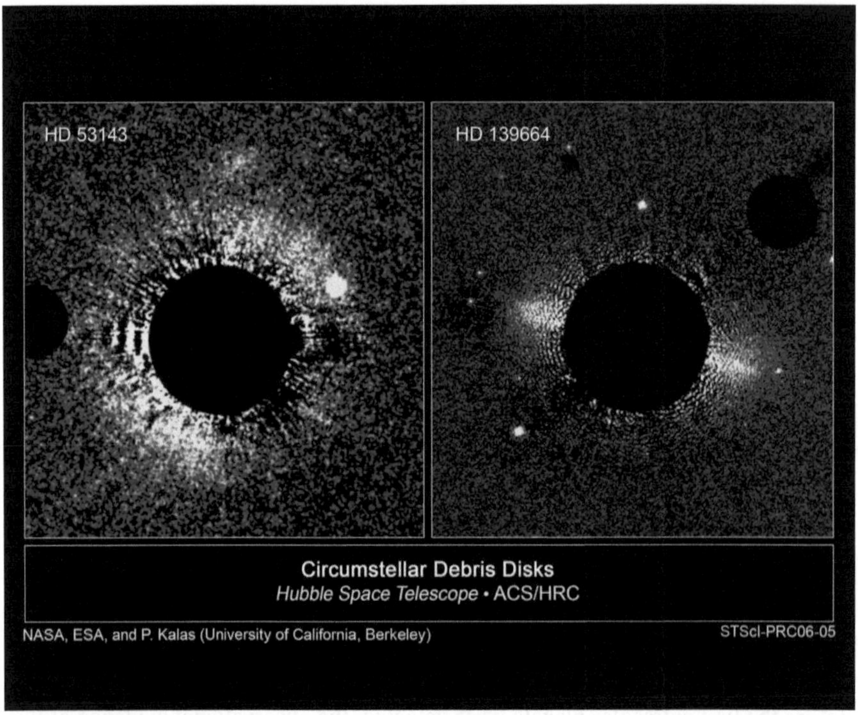

Fig. 6.6 Dust debris around HD 53143 and HD 139664

- Toroidal OH outflow geometries observed in VLBI[3] are experiments,
- Coexistence of water-ice and water-vapor at distances of several 100 AU,
- Dust and complex organics located at the water condensation radius of M giants and supergiants.

The presence of water in these outflows is naturally explained by the destruction of volatile rich components that are expected in an Oort cloud like structure.

The effect of heating and resultant radiation of any orbiting dust cloud particles around red giants should be also detectable in the IR. The Kuiper belt objects (KBO) start to sublimate because they are heated up to 170 K. This leads to an IR excess at 25 μm.

Using IRAS[4] data, Jura, 2005 [176] searched for IR-signatures. With the Spitzer telescope the IR excess was measured around first ascent red giants. The data showed that this radiation seems to be caused by interstellar cirrus and not by KBO like objects. If there is any KBO like structure around these stars, then it must be less massive than the Solar System's.

In principle an IR excess among red giant stars could be explained by

- Dust that is ejected from the stars,
- Interstellar dust that is nearby the star is heated ("cirrus hot spot"),
- Disintegration of comets.

Sublimating comet disks are discussed in the paper of Mates, Whitmire and Reynolds, 1989 [219].

Properties of debris dust around solar-type stars were given by Carpenter et al., 2009 [57]. In total 314 were investigated with the Spitzer telescope. For stars younger than 300 million years 15% showed a photospheric IR (at 24 μm) greater than 10.2% and this value falls to zero for older stars. A Spitzer study of debris disks in the young nearby cluster NGC 2232 was made by Currie, Plavchan and Kenyon, 2008 [88]. The frequency of debris disks around A-type stars appears to increase from ∼25% at 5 Myr to ∼50%–60% at 20–25 Myr. Older A-type stars have smaller debris disk frequencies: ∼20% at 50–100 Myr. This findings suggest that the frequency of icy planet formation is between 0.5–0.6. Thus, most A-type stars (∼1.5–3 $M_{\odot}$) produce icy planets.

6.3.2 Dust Around Vega

Vega (αLyr) is a very well known star, the brightest star in the northern sky hemisphere. It is located at a distance of about 25 Ly and it belongs to the group of A0 stars with a surface temperature of about 10 000 K. In comparison to the Sun, Vega

[3] VLBI means very long basement interferometry. Signals from radiotelescope from extremely distant stations on Earth are brought to interfere.

[4] Infrared Astronomical Satellite.

is relatively young, its age is about 350 million years. It rotates very fast (rotation period 12.5 h, mean rotation for the Sun: 25.4 days), and is therefore flattened; because of this effect this star is about 2000 K cooler at the equator than at the poles. This effect is also known as gravity cooling.

Observations with the Infrared Astronomy Satellite in 1983 revealed that there are dust particles around Vega (in fact that was the first observation of dust around another star) (Aumann et al., 1984 [13]). Observations at mm wavelengths are favorable for the detection of dust particles since at that wavelengths, the contrast between the star and the dust is much smaller than e.g. in the optical. In 1998 observations revealed structure in the dust cloud around Vega that can be only explained by the presence of a planet. The observations of Vega are very well suited to study the dust because from the Earth Vega is seen nearly pole on. The high resolutionobservations revealed two prominent peaks of dust emission, one located offset 60 AU to the southwest of Vega, the second at 75 AU to the northeast. A planet in an eccentric orbit may cause these offsets. From observations at 25, 60 and 100 μm it was concluded that the dust cloud extends from 41 to 177 AU (Buitrago and Mediavilla, 1985 [50]).

Marsh et al., 2006 [217] made observations at the wavelengths 350 and 450 μm with the SHARC II camera at Caltech Submillimeter Observatory. The resolution of the image was 9.7 and 11.1 arcsec respectively. The observations are consistent with a characteristic grain size of about 1 mm. Observations at 850 μm are compatible with a cm grain size and much smaller grains can be observed by observations e.g. at 70 μm. They concluded the existence of a Neptune sized planet at 65 AU.

Vega was also observed with the Spitzer IR telescope (Fig. 6.7).

Manoj and Bhatt, 2005 [215] investigated Vega like stars and their relation with dust disks. As parameter for the size and mass of the dust disks the fraction L_{dust}/L_{star} was estimated. They found that debris disks seem to survive longer

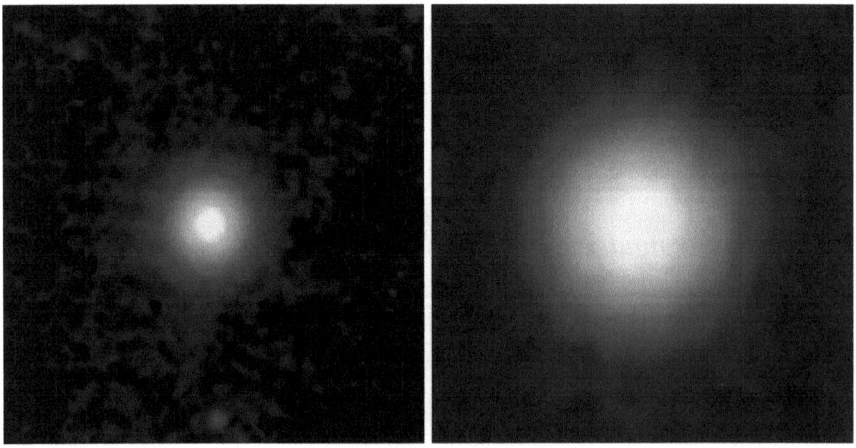

Fig. 6.7 Dust debris around Vega. Observations were made by the Spitzer IR telescope, *left* at 24 microns, *right* at 70 microns where the dust becomes more conspicuous. Courtesy: JPL, Spitzer IR telescope

around later type stars as compared to early type stars and the fractional dust luminosity decreases steadily with increasing stellar age. As an indicator for the age of the stars, velocity dispersion measurements from the HIPPARCOS satellite were used. To explain a velocity dispersion correlation with stellar age two mechanisms are possible: (i) stellar two-body encounters (very unlikely because of the large stellar distances) and (ii) interactions of stars with massive interstellar clouds.

6.4 Water Detection on Extrasolar Planets

The detection of atmospheres on extrasolar planet is a very difficult task. 71% of the Earth is covered by oceans but up to now it is the only planet with water in liquid form on its surface. Venus might have had water on its early history, on Mars water may exist in a frozen state near the surface and climatic changes have occurred and formed river-like structures that are observed on its surface. There exists the possibility to find condensed water in the atmospheres of Jupiter and Saturn and in deeper layers of Uranus and Neptune. Subsurface oceans may exist on several satellites of the giant planets. But how can we detect water on extrasolar planets, how can we detect whether these objects have even an atmosphere?

6.4.1 Detection of Planetary Atmospheres

For the detection of extrasolar planetary atmospheres we are at the moment in a state similar to that 200 years ago when the first attempts were made to study atmospheres of planets in our solar system. Generally, the effects of planetary atmospheres are:

- Refraction: to an observer on the Earth, any object remains visible above the horizon even when its actual position is already below the horizon. This causes a bending of light in a planetary atmosphere.
- Detection of clouds: the effect of clouds in the spectra are mainly a smoothing of features and changing the reflection and albedo. If the planet is covered densely by clouds then its optical albedo is usually high, if there are no clouds, the atmosphere absorbs most of radiation.
- Clouds can also deplete the atmosphere of abundant elements (e.g. Al, Si, Mg, Ca and Fe) by rainout.
- Condensed particles in a planetary atmosphere: the species of such condensed particles depends on the effective temperature of the atmosphere (which is related to incoming amount of stellar radiation).
  - Low T_{eff}: ices like water ice or ammonia ice.
  - High T_{eff}: grains of iron or silicates.
- Extrasolar planetary atmospheres can be detected by an additional absorption during transits (Fig. 6.8).

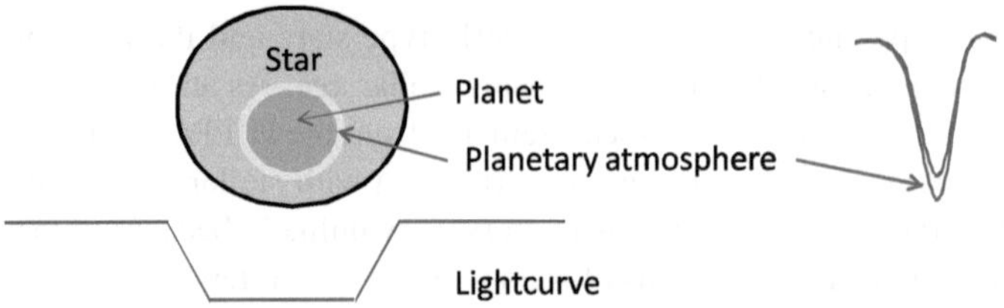

Fig. 6.8 Effect of planetary atmosphere during a transit event on the absorption line. Also the dip in the lightcurve may become more pronounced because of a dense atmosphere

The dip in the absorption lines during a planetary transit is very small. Seager and Sasselov, 2000 [302] predict the possibility to observe a relative decrease in flux of about 10^{-4}. In order to detect such small variation strong planet spectral features have to be observed and several absorption frequencies (NaI, KI, HeI) are proposed for observations.

Let us consider the transit of a planet. The stellar flux will be reduced by the amount of the ratio of the area of the planet to the star. If the planet has an atmosphere, then some flux from the star will pass through the optically thin part of the planet's atmosphere. In the case of stellar occultations by giant planets in the solar system, the limb of the giant planets is either defined by the cloud tops or by the 1 bar level.

Assuming that the radius of the star is $R_* = 1.3\ \mathrm{R}_\odot$ and the radius of the planet is $R_P = 1.54\ \mathrm{R}_J$, where R_J denotes the radius of Jupiter. The ratio of planet-to star area is on the order of 10^{-4} to 10^{-3}. For the object HD 209458 an estimate for the limb radial depth of $0.01\ R_P$–$0.05\ R_P$ can be assumed. The planet's absorption features are superimposed on the stellar flux. In order to detect these small features, they must be optically thick. In an isothermal atmosphere with gas scale height H, the angle of refraction for a ray passing at planetocentric distance r is given by:

$$\theta = \nu(r)\sqrt{\frac{2\pi R_P}{H}} \tag{6.4}$$

$\nu(r) = n - 1$ is the atmospheric refractivity at r. Assuming a H_2, He mixture, $\nu(r) \sim 1.2 \times 10^{-4}$. The index of refraction, n, is a function of wavelength λ and r.

6.4.2 Hot Jupiters

The first discovery of a massive planet, 51 Peg B, was made in 1995 by Mayor and Queloz, 1995 [221]. This object is also close to its parent star 51 Peg, at a distance of only 0.05 AU. A new class of planets was defined when it became clear that such objects are quite common. About 50% of known extrasolar planets belong to this group. These "hot Jupiters" have the following characteristics:

- orbital distance from their parent star is less than 0.2 AU,
- high effective surface temperatures larger than 1000 K.

A study of a typical hot Jupiter (HD 209458 b, see Chap. 7) was made to model the expected dynamic processes in its atmosphere. The day-night temperature differences are expected about 200 K. Also the existence of jets and up- and downwelling regions is predicted (Showman and Guillot, 2000 [309]).

The spectroscopic signatures of transiting hot Jupiters were studied by Brown et al., 2000 [46]. The depth of absorption features could reach up to 10^{-3} of the parent star's continuum intensity.

Due to the close distances from their host stars, the temperatures in their atmosphere are expected to be high. The escaping molecules and their dissociation and ionization products can form a large envelope around such a planet. Evaporation of such an atmosphere is caused by

- Thermal escape, Jeans escape; if we assume that the individual molecules are Maxwellian distributed, then molecules from the high tail of this distribution can escape, if at a given level in the atmosphere the mean free path is comparable to the scale height. According to Maxwell's distribution, the kinetic energy of the molecules depends on the mass and velocity (which depends on the temperature in the atmosphere):

$$E_{\mathrm{kin}} = \frac{1}{2}mv^2 = \frac{3}{2}kT \tag{6.5}$$

  where $k = 1.3806504(24) \times 10^{-23}$ J K^{-1}. The more massive a gas molecule, the lower the average speed at a given temperature, it is less likely to escape; e.g. hydrogen escapes more easily than carbon dioxide. If a planet has a higher mass, the escape velocity, v_{esc} is greater, and fewer particles escape.

$$v_{\mathrm{esc}} = \sqrt{\frac{2GM}{R}} \tag{6.6}$$

  where M is the mass of a planet, R its radius.
- Non-thermal evaporation: these processes are caused by the stellar wind and UV flux. If there is no strong planetary magnetic field, solar wind and EUV flux contribute to the escape of heavier particles.
- Gravitational transfer to the massive companion.

Assuming an energy input of 10^{18} W as particle flux on 51 Peg b, molecules such as H_2O, CO, CH_4, N_2 could be teared away. These become quickly ionized and fill a space around the planet. Since the planet is assumed non magnetized, a comet-like tail forms and no torus. This structure should be observable during a stellar occultation (in the IR from 1.5 to 2.5 µm). The radius of the observable exosphere should be about 10 times the radius of the planet (for more details see Coustenis et al., 1997 [81]).

Moutou et al., 2001 [238] made observations of HD 209458 b with UVES at the VLT covering most of the 328–669 nm range. They were not able to detect HD 209458 b's exosphere at a level of 1%, a value close to the predictions.

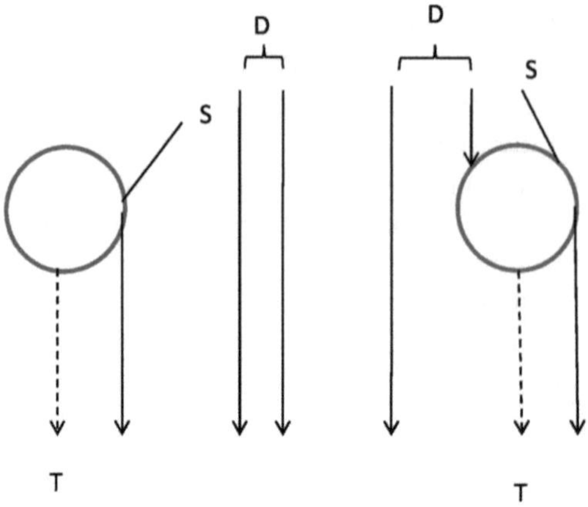

Fig. 6.9 Three sources of radiation come to an observer from the star/planet system. Direct radiation from the star (D) is partially blocked during the transit. Thermal radiation from the planet (T) varies because of presumed temperature and emissivity variations across the planet's surface. The scattered component (S) varies because of the changing illuminated phase and because of possible scattering-angle dependences caused by particles in the planetary atmosphere. Note that, because of the angular extent of the parent star, significantly more than a hemisphere of the planet is partially illuminated (adapted from [49])

Brown, 2001 [49] investigated features on the spectra of transiting extrasolar giant planets. A spectrum ratio $R(\lambda)$ is defined. This is the ratio of spectra taken in and out of transit.

$$R(\lambda) = \frac{F_{\text{transit}}(\lambda)}{F_0(\lambda)} \tag{6.7}$$

For close extrasolar giant planets line depths of 10^{-3} relative to the stellar continuum can occur. The most useful diagnostics are likely to be the near-infrared bands of molecules like H_2, CO, H_2O, CH_4, and the visible/near-IR resonance lines of the alkali metals.

As indicated in Fig. 6.9 there are three ways for light from the star/planet system to reach the observer. (i) Most of the light is direct emission from the star. (ii) Thermal emission from the planet, (iii) light from the star that is scattered into the line of sight by the planet. These contributions can be written as:

$$R = \frac{F_0 + \delta F}{F_0} = 1 + \frac{\delta F_{\text{direct}} + \delta F_{\text{therm}} + \delta F_{\text{scat}}}{F_0} \tag{6.8}$$

Note that all of these contributions are dependent on the wavelength λ. The observed flux from a star can change either by intrinsic variations due to stellar activity or by transiting planets that occult a small part of the star. Spectral lines from a star show a center to limb variation. This is explained by the theory of line formation. Different lines are formed at different depths in the stellar atmosphere. As a crude rule we may state that the deeper the line, the higher it is formed in the atmosphere. If a planet passes in front of a star it obscures only a specific part of the star's visible surface. The variation of line profiles across the disc is several tens of a percent. Therefore, the spectrum as a whole may vary by several 0.1%. Making a time average over the transit, this effect is reduced but not eliminated. From these considerations it follows, that it is necessary to model the center-to-limb variation of spectral lines in order to extract signatures coming from the planet's atmosphere itself. We can write for the wavelength-dependent obscuration magnitude $(\delta A/A)_{\mathrm{atm}}$ caused by the atmosphere of a planet:

$$\left(\frac{\delta A}{A}\right)_{\mathrm{atm}} = \frac{2\pi r_p (kT/(g\mu)}{\pi r_*^2} \tag{6.9}$$

T is the temperature, g the surface gravity, μ the mean molecular weight. The scale height for an atmosphere of an object with $g = 10^3\ \mathrm{cm\,s^{-2}}$, $T = 1440$ K, molecular H, $r_p = 1.55\ \mathrm{r_{Jup}}$, $r_* = 1.3\ \mathrm{r_\odot}$ is $H = 770$ km, and for a Sun-sized star:

$$\left(\frac{\delta A}{A}\right)_{\mathrm{atm}} = 2 \times 10^{-4} \tag{6.10}$$

Further details of these calculation are found in Brown, 2001 [49].

Giant planets at orbital distances smaller than 0.1 AU from their stars are rapidly synchronized by tidal interactions. Departures from synchronous rotation can occur because of the strong gravitational interaction fluid-dynamical torques within such planets may occur. Showman and Guillot, 2002 [309] calculated characteristic properties of such planetary atmospheres:

- day-night temperature difference: ~500 K,
- characteristic wind speeds: ~2 $\mathrm{km\,s^{-1}}$,
- clouds could exist predominantly either on the dayside or on the nightside, depending on the circulation model.

These models were applied on the 51 Peg b planet.

Atmospheric lensing and oblateness effects during an extrasolar planetary transit were investigated by Hui and Seager, 2002 [168]. Planet oblateness induces an asymmetry to the transit light curve about the point of minimum flux. If θ_S, θ_I denote the source and image positions (see Fig. 6.10), θ_D the angle of deflection and

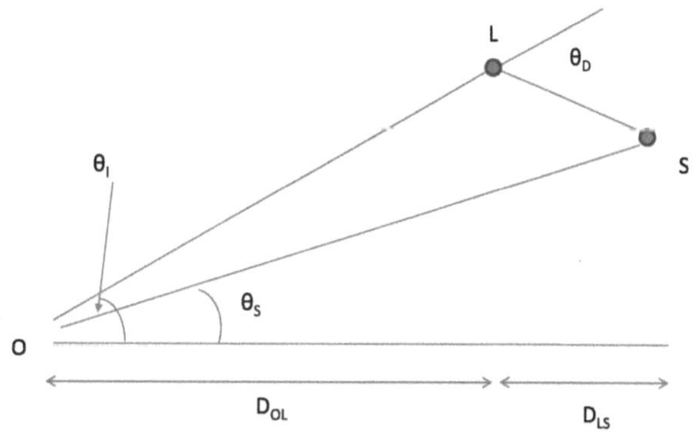

Fig. 6.10 The deflection is exaggerated, the diagram is not to scale. The light ray is the line joining O (observer), L (lens) and source S

D_{LS} the distance between the lens and the source and D_{OL} the distance between the lens and the observer, then:

$$\theta_S = \theta_I + \theta_D \frac{D_{LS}}{D_{OL} + D_{LS}} \tag{6.11}$$

The observed flux from a star during an extrasolar planet transit is given by:

$$F(t) = \int d\theta_S \sum AI[\theta_S - \theta_*(t)]W(\theta_I) \tag{6.12}$$

and the function $W(\theta_I)$ describes occultation and absorption.

Extrasolar giant planets that are close to their parent star should exhibit a complex circulation pattern. The large spatial scales of moving circulation structures on HD 209458 b may generate detectable variability of the planet's atmospheric signatures (Menou et al., 2003 [229]).

6.4.3 Water on Extrasolar Planets

The Object HD 209458 b was detected in 1999 (see e.g. Rauer et al., 2004 [272]). Many of extrasolarplanets have been detected by measuring the changes of a star's radial velocity due to the gravitational attraction by the presence of a planet or several planets. HD 209458 b was the first object that was confirmed by a transit measurement, when the planet passes in front of its parent star, a dip of 1.7% in the star's brightness was recorded. The distance of this object is 153 Ly from Earth. Its mass has been determined to be about 63% the mass of Jupiter, thus it belongs to the group of giant planets. However, it is at a distance of only 7×10^6 km to its parent star which is 100 times closer than Jupiter to our Sun. Few years after its discovery, oxygen and carbon were found in the planet's atmosphere. Because of its small distance to its solar like parent star, the planet is assumed to lose continuously volatiles, it is estimated that it loses up to 10^4 tons per second. Data with the Hubble Space telescope revealed the presence of water absorption. Hydrogen escapes and extends away from the planet which probably developed a cometary tail like structure. Due to its close distance to its parent star, HD 209458 b circles the star every 3.5 days. The expected temperature of its atmosphere is about 10 000°C, therefore hydrogen expands into space and then extreme tidal forces help to release this gas completely from the planet's gravitational field. In addition to that, a strong stellar wind of charged particles contributes to a stripping of hydrogen. By these processes the whole atmosphere will be blown off leaving behind a core (solid or liquid) with about 10 times the Earth mass. An artist's impression of this evaporating planet is given in Fig. 6.11.

6.4.4 Some Model Calculations

In this subsection we present the work of Menou et al., 2003 [229] who simulated the motion of a thin, homogeneous layer of hydrostatically balanced, inviscid fluid

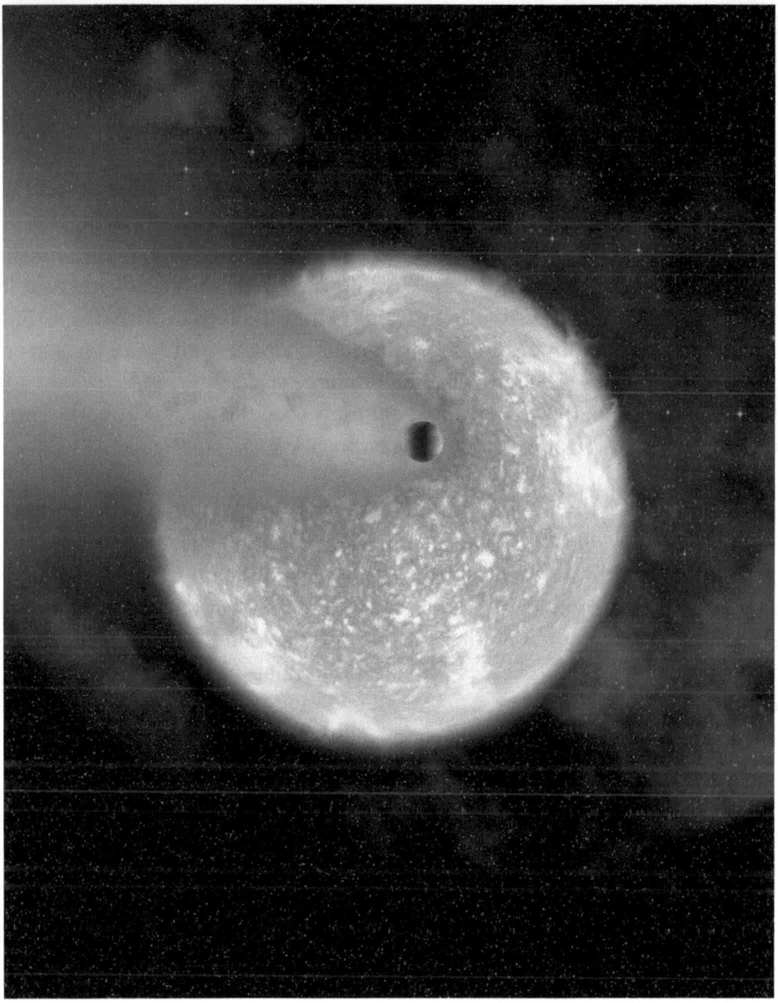

Fig. 6.11 Water claimed in Evaporating Planet HD 209458 b Illustration Credit: European Space Agency, Alfred Vidal-Madjar (Institut d'Astrophysique de Paris, CNRS), NASA

with a free surface. Such a fluid is subject to gravitational and Coriolis forces. $\mathbf{v}$ denotes the horizontal velocity, h the thickness of the modeled layer, $f = 2\Omega \sin \phi$ is the Coriolis parameter, Ω is the rotation rate, ϕ the latitude, g the gravitational acceleration and $\mathbf{k}$ is the unit vector normal to the surface of the planet

$$\frac{\partial \mathbf{v}}{\partial t} + \mathbf{v} \cdot \nabla \mathbf{v} = -g \nabla h - f \mathbf{k} \times \mathbf{v} \tag{6.13}$$

$$\frac{\partial h}{\partial t} + \mathbf{v} \cdot \nabla h = -h \nabla \cdot \mathbf{v} \tag{6.14}$$

In dimensionless form, shallow-water equations become functions of the Rossby (Ro) and Burger (Bu) numbers:

$$Ro = \frac{U}{|f|L}, \qquad Bu = \left(\frac{L_D}{L}\right)^2, \qquad L_D = \sqrt{gH}|f| \tag{6.15}$$

U, L, H are the characteristic velocity, length and layer thickness scales. L_D is the Rossby deformation radius. The Rossby number measures the importance of rotation on the flow, the Burger number measures the stratification in the atmosphere via the Brunt-Väisälä frequency. As it is known, the atmospheric structure in the giant planets in the solar system is in bands. This structure evolves from shallow-water turbulence. Two dimensional turbulence is characterized by an inverse energy cascade—this means a transfer from small to large scales.

A simple atmospheric model must take into account several effects: The photodissociation rate $J(z)$ at height z in the atmosphere is

$$J(z) = \int \sigma_\nu F_\nu e^{-\tau_\nu} d\nu, \qquad \tau_\nu(z) = \int_u^{z_{\max}} \kappa_\nu ds \tag{6.16}$$

F_ν is the input flux from the host star,[5] τ_ν the optical depth which depends on the opacity (cm^{-1}) and σ_ν is the photodissociation cross-section. All these quantities depend on the frequency ν. Photons may be absorbed and Rayleigh scattering could also be important. Both processes provide a shielding of an atmosphere from UV radiation.

The temperature profile of a planet depends on the optical depth:

$$T^4 = \frac{3}{4} T_{\text{eff}}^4 \left(\tau + \frac{2}{3}\right) + \mu_0 T_{\text{eq}}^4 \left(1 + \frac{3}{2}\left[\frac{\mu_0}{\gamma}\right]^2 - \frac{3}{2}\left[\frac{\mu_0}{\gamma}\right]^3 \ln\left[1 + \frac{\gamma}{\mu_0}\right] - \frac{3}{4}\frac{\mu_0}{\gamma} e^{-\gamma\tau/\mu_0}\right) \tag{6.17}$$

μ_0 is the cosine of the angle between each point on the surface of the planet and the illuminating host star. T_{eq} is the planet's equilibrium temperature, T_{eff} stands for additional internal heating, $T^4 = T_{\text{eff}}^4 + T_{\text{eq}}^4$ at optical depth 2/3 and $\mu_0 = 1$, γ is the ratio of the optical depth of absorption to that of emitted radiation. This value is characteristic for the deposition of stellar energy into a planetary atmosphere. Miller-Ricci et al., 2009 [230] modeled the atmosphere of the super Earth Gliese 581. For an albedo of 0.15 (cloud–free model) a temperature of $T_{\text{eq}} = 368$ K was found and for highly reflective sulfuric acid clouds $T_{\text{eq}} = 295$ K. For T_{eff} a value of 50 K was assumed.

The pressure structure can be deduced from integrating the equation of hydrostatic equilibrium:

$$\frac{dP}{d\tau} = \frac{g}{\kappa_{\text{gr}}} \tag{6.18}$$

[5]The number of photons $cm^{-2}\,s^{-1}\,Hz^{-1}$.

κ is the Planck mean opacity. In order to check wether the atmosphere is stable to convection, the stability criterion to be applied is:

$$\frac{d \ln T}{d \ln P} < \nabla_{\mathrm{ad}} \tag{6.19}$$

The adiabatic gradient $\nabla_{\mathrm{ad}} = (\Gamma - 1)/\Gamma$, Γ is the adiabatic index. If this criterion is satisfied, the radiative temperature-pressure profile can be obtained; if not, the pressure-temperature profile follows an adiabate.

From the obtained temperature pressure profile, the emergent intensity can be calculated from:

$$I(\lambda, \mu) = \frac{1}{\mu} \int_0^\tau S(T) e^{-\tau'/\mu} d\tau' \tag{6.20}$$

In this expression S is the source function which can be approximated by a Planck function. Scattering in an atmosphere strongly depends on wavelengths, being stronger at shorter wavelengths (e.g. Rayleigh scattering $\sim\lambda^{-4}$), therefore, especially in the IR it can be neglected.

Atmosphere and Water on Exoplanets

Atmospheres of extraterrestrial planets can be directly studied during a transit. Two transit events with the Near Infrared Camera and Multi Object Spectrograph (NICMOS) camera on the Hubble Space Telescope (HST) were observed for the object GJ 436 b. In order to detect the atmosphere, high-cadence time series of prism spectra covering the 1.1–1.9 μm spectral range were analyzed (Pont et al., 2009 [266]). This object is an extrasolar hot Neptune. The authors measured a flat transmission spectrum at the level of a few parts per 10 000 in flux, with no significant signal in the 1.4 μm water band.

During the impact of comet Shoemaker Levy on Jupiter (1994), a 22 GHz water MASER emission line was observed. This line can be used as a diagnostic for the existence of water on exoplanets. An observational program (called ITASEL = Italian Search for Extraterrestrial Life) started 1999 using the 32 m dish of the Medicina radiotelescope and from 32 candidates four emissions can be suspected. If such observations can be definitely confirmed it will prove the importance of cometary impacts for the deposit of water on exoplanets (Cosmovici et al., 2006 [80]).

Planets with oceans will appear brighter (fainter) at small (large) phase angles than planets with diffuse-scattering surfaces. In addition, such planets will polarize incident starlight by us much as 70% under idealized conditions. This was studied in detail by Williams and Zugger, 2010 [358]. They conclude that some exoplanets with thin atmospheres will unambiguously show evidence of water (or not) from polarization of light reflected by the surface. With the Spitzer telescope a search for Water in the transiting exoplanet HD 189733 b was made. It was shown that current observational capabilities are insufficient for detecting water vapor (Ehrenreich et al., 2007 [115]).

6.4.5 Super Earth Planets

Super-Earth planets are planets with masses between 1 and 10 Earth masses. Such masses are still beyond the mass of the gas giants, e.g. the mass of Neptune is 17 M_{Earth}. The atmospheres of such objects may be different from atmospheres for typical terrestrial planets. Some of them may possess hydrogen rich atmospheres. These objects should show strong H_2O features in their spectra. Objects that have lost most of their hydrogen could exhibit strong CO_2 features. Generally, this type of objects seems to represent a transition between rocky objects and Neptune-like planets. The discovery of super-Earths was discussed e.g. in Elkins-Tanton and Seager, 2008 [120]. They argue that there are three ways how such objects may obtain atmospheres:

- capture from the primordial nebula from which they were formed,
- degassing during accretion,
- degassing from tectonic activity during their evolution.

The habitability of super earths was discussed e.g. by Kaltenegger and Kasting, 2008 [177].

As we have mentioned in the chapter about meteorites, chondrites are the most primitive meteorite class. Hydrogen and oxygen are present mainly in the form of OH, the water content in the achondrites is lower (up to 3%). The degassing during accretion leads to a wide range of the mass of their possible atmospheres from less than 1% of the planet's total mass to 6% by mass of hydrogen, 20% of water and 5% of carbon compounds. Planets with deep surface liquid water oceans could have formed after the accretion has stopped.

It is generally believed that super-earths may have a large variety of atmospheres. Some of them may even retain a hydrogen rich atmosphere (Miller-Ricci, Seager and Sasselov, 2009 [230]). Super-Earths with massive hydrogen atmospheres will reveal strong spectral features due to water, whereas those that have lost most of their hydrogen (and have no liquid ocean) will be marked by CO_2 features and a lack of H_2O.

The formation and tidal evolution of hot super-Earths in multiple planetary systems was reviewed by Zhou, 2009 [373]. These objects are exoplanets with masses <10 Earth masses and orbital periods <20 days (like e.g. HD 40307 and GJ 436). Around 8 hot super-Earths have been discovered in the neighborhood of solar system. Mayor et al., 2009 [222] reported on the discovery of a planetary system with 3 super-Earths (4.2, 6.9, and 9.2 M_{Earth}) orbiting HD 40307. The detection was made with the HARPS instrument on the ESO 3.6 m telescope by radial velocity measurements.[6] The first exoplanet observations from the Gran Telescopio Canarias (GTC) using the OSIRIS tunable filter imager were presented by Ford et al., 2010 [133]. The authors show that transiting super Earth planets can be detected at near IR wavelengths and such follow up observations with high photometric precision

[6]HARPS: High Accuracy Radial velocity Planetary Search project.

are extremely valuable for interpretation of space observations. Observations carried out with HARPS at the 3.6 m telescope at La Silla on the star CoRoT-7 were made by Queloz et al., 2009 [269]. Two objects, Corot-7b and Corot-7c with 4.8 and 8.4 Earth masses were found. Corot-7b has a density that is comparable to that of the Earth.

At least three planets with masses of 5.1, 18.2, and 24.0 M_{Earth} seem to orbit the star 61 Vir which is of spectral type G5 V. This was deduced from 4.6 years of combined Keck/HIRES and Anglo-Australian Telescope precision radial velocities (Vogt et al., 2010 [347]).

M-dwarf stars have small radii. A planet with 2 Earth radii orbiting around such a star in the habitable zone (here defined as the region were liquid surface water may exist on a planet) will produce a 0.5% drop of luminosity of the star during transit and can be detected even with small aperture telescopes (Irwin et al., 2009 [170]).

Chapter 7
Water in Interstellar Space and Stars

In this chapter we discuss the occurrence of water in interstellar space, especially in star forming regions. Water molecules were even detected in the spectra of sunspots.

7.1 Interstellar Medium

The space between stars is not empty but filled with a gas and dust component that is called the interstellar medium. Stars are born out of interstellar clouds and at the end of their evolution they enrich the interstellar medium with elements heavier than He. Depending on their masses, stars become either unstable emitting strong stellar winds or expelling their outer atmosphere, the more massive stars explode and become a supernova.

Elements heavier than helium were not produced during the Big Bang, these elements have been synthesized by nuclear fusion in the stellar interior and are important for the origin of planetary systems and the origin of life itself.

7.1.1 Physical Properties

The Sun and the solar system is located at a distance of about 30 000 Ly from the center of the galaxy. The density of the interstellar medium in the solar neighborhood is about 0.25 particles per cm^3.[1] Approximately 99% of the interstellar medium is composed of interstellar gas, and of its mass, about 75% is in the form of hydrogen (either molecular or atomic), with the remaining 25% as helium. The interstellar gas consists partly of neutral atoms and molecules, as well as charged particles, such as ions and electrons.

The gas can be observed as

[1]For comparison: the air we breathe contains about 10^{19} molecules per cm^3.

A. Hanslmeier, *Water in the Universe,* Astrophysics and Space Science Library 368,
DOI 10.1007/978-90-481-9984-6_7,

- Cold clouds of neutral atomic or molecular hydrogen; such clouds can be detected at a wavelength of 21 cm in the radio part of the electromagnetic spectrum. This wavelength (corresponding to a frequency of 1420 MHz) is emitted when the spin of the electron of a neutral hydrogen atom changes from its direction parallel to antiparallel to the spin of the proton (hyperfinestructure).
  Such regions are also called H-I regions. The temperature is below 100 K.
- Hot ionized hydrogen clouds near hot young stars, also called H-II regions. These objects can be seen as bright emission nebulae mainly in the red light of Hydrogen-Alpha, Hα. This line is emitted when an electron transition from the level 3 to 2 occurs in the hydrogen atom. The temperatures can reach values up to 10^6 K.

Generally, the interstellar medium is very dilute, even in clouds densities range from a few hundred up to about 10^8 particles cm^{-3}. In total the interstellar medium contributes about 15% percent to the visible mass of a typical galaxy.

The cold clouds of neutral or molecular hydrogen are important for the formation of new stars. They can become gravitationally unstable as soon as their density is above some critical value, the so called Jeans mass. Then a collapse of the cloud and subsequent fragmentation starts and new stars are born in a cluster.

Bright emission nebulae (ionized H) are formed when a bright young star emits large amounts of UV radiation that ionizes the surrounding gas clouds. The hydrogen atoms become ionized and when they recombine, mostly the red hydrogen line Hα is emitted. There are many famous examples of such objects like the Orion Nebula (Fig. 7.1) or the Trifid Nebula. In these objects young stars are found (star forming regions).

The interstellar dustcomponent is much more difficult to describe by physical parameters since e.g. no spectrum is observed. Some dust clouds can be even seen with the naked eye: certain parts in the Milky Way appear dark due to the absorption by the dust particles. Generally dust causes:

- Absorption (see Fig. 7.2),
- Reddening of starlight,
- Polarization, this polarization can be explained by an alignment of the non spherical dust particles due to the galactic magnetic field.

As an example we give the Horsehead Nebula (Fig. 7.2). The darker parts of the Horsehead are caused by absorption due to thick dust. The lower part of the Horsehead's neck casts a shadow to the left. Streams of gas leaving the nebula are funneled by a strong magnetic field. Bright spots in the Horsehead Nebula's base are young stars just in the process of formation.

The interstellar environment of our galaxy was reviewed by Ferrièere, 2001 [132].

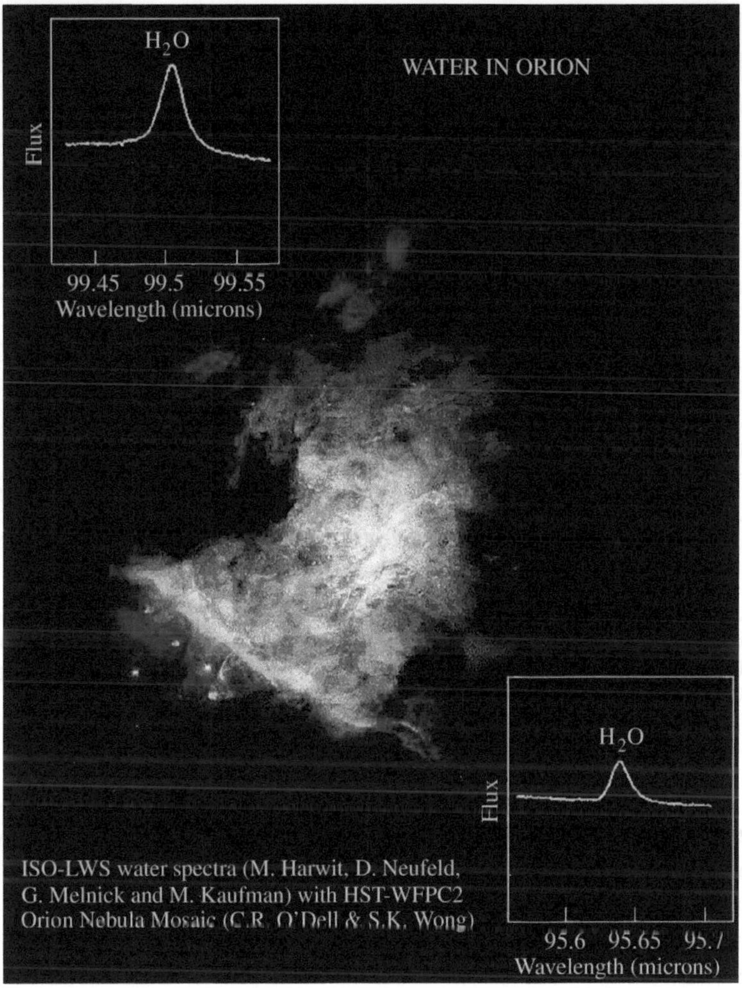

Fig. 7.1 Spectral signatures of water in the Orion nebula, M42. The distance of this famousobject is about 1 500 Ly. The image is a mosaic of more than 40 individual Hubble Space Telescope (HST) images. Credit: ESA/NASA

7.1.2 Molecules in the Interstellar Medium

The spectral features of molecules are different from those of atoms as well as their transition mechanisms. For atoms the electrons jump between different discrete energy levels. The spectral features of molecules are generated by different energy levels or by rotation and vibration transitions. This will be explained in the Appendix.

The first such molecule detected in the interstellar medium was the methylidyne radical (CH) in 1937. While Saha, 1937 [289] proposed the identification of some

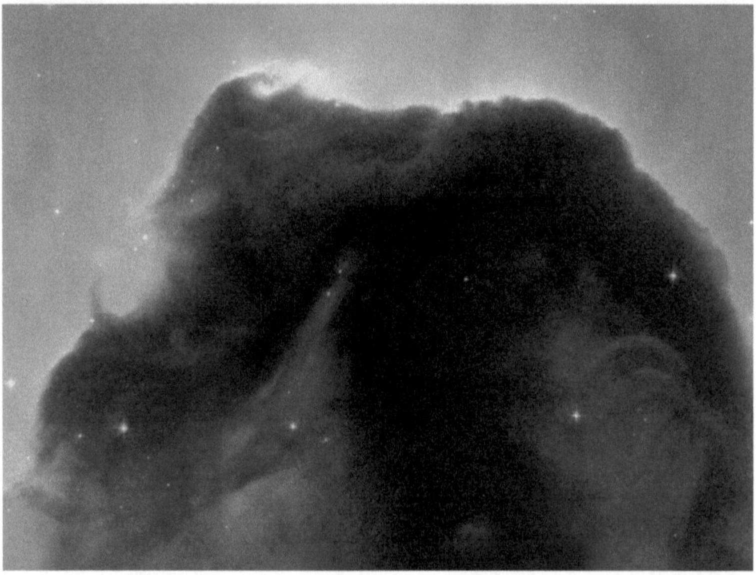

Fig. 7.2 The so called horse head nebula. This object is also part of the giant Orion molecular and dust complex at a distance of 1 500 Ly. Image Credit: NASA, NOAO, ESA and The Hubble Heritage Team (STScI/AURA). Acknowledgment: K. Noll (Hubble Heritage PI/STScI), C. Luginbuhl (USNO), F. Hamilton (Hubble Heritage/STScI)

observed spectral lines to atoms, Swings and Rosenfeld, 1937 discussed the possibility to find lines of molecules such as CH, OH, NH, CN and others (Swings, Rosenfeld, 1937 [331]).

The hydroxyl radical OH was the first interstellar molecule detected at radio frequencies (Weinreb et al., 1963 [356]).

Interstellar molecules are formed by chemical reactions within interstellar or circumstellar clouds of dust and gas. By interaction with a cosmic ray particle, an atom becomes ionized. Such a positively-charged molecule attracts a nearby reactant (electrostatic force). Molecules can also be generated by reactions between neutral atoms and molecules, although this process is generally slower.

In interstellar clouds a huge amount of different molecules can be found including also complex organic molecules.

Interstellar molecules[2] can be observed at radio-, millimeter-, sub-millimeter and IR-wavelengths. From these observations many organic molecules were detected. It must be stressed that their abundance relative to hydrogen (molecular) is less than 10^{-4} and in many cases even less. However, even very complex molecules like $HC_{11}N$ and diethyl ether, CH_3–CH_2–O–CH_2–CH_3 have been detected.

[2]See also: D. Buhl, Chemical Constituents of Interstellar Clouds, 1971, Nature 234, 332.

Table 7.1 Some important molecules detected in the interstellar medium

CO	Carbon monoxide
FeO	Iron (I) oxide
H_2	Molecular hydrogen
HO	Hydroxyl radical
NO	Nitric oxide
NaCl	Sodium chloride
C_2H	Ethynyl radical
CO_2	Carbon dioxide
H_2C	Methylene
H_2O	Water
H_2S	Hydrogen sulfide
NH_2	Amino radical
SO_2	Sulfur dioxide
C_2H_2	Acetylene
H_2CO	Formaldehyde
NH_3	Ammonia
CH_4	Methane
HCOOH	Formic acid
CH_3OH	Methanol

In the cold parts of the interstellar medium there are icy dust grains.[3] These particles are important because they may act as catalysts.

The interstellar medium may also appear as dark clouds that absorb starlight. The densities are several 10^2 particles cm^{-3}. In such clouds polycyclic aromatic hydrocarbons (PAHs), fullerenes, carbon-chains, diamonds etc. are found.

Some of the molecules that have been detected in the interstellar medium are listed in Table 7.1. This table is only an incomplete sample of the long list of molecules that have been detected.

7.1.3 Interstellar Dust Lifecycle

It is believed that pristine water is formed and grows on the surface of icy dust grains in dark interstellar clouds. This has been proven experimentally by atomic beams of oxygen and deuterium aimed at a porous water ice substrate (H_2O) held at 10 K (Dulieu et al., 2010 [110]). Dust is observed to form in the ejecta from red giants, supergiants, novae, and supernovae through nucleation, chemical growth, and coagulation. in interstellar clouds, dust grows icy mantles, consisting of simple ices such as H_2O, CH_3OH, CO, CO_2, and CH_4, through accretion and reaction of gas phase

[3]This has been investigated with the ISO satellite, Infrared Space Observatory.

species. There is a balance between formation and destruction (e.g. by sputtering in supernova shocks). Interstellar depletion can have different causes: stardust injection, accretion in clouds, dust destruction in interstellar shocks. Si bearing dust has a lifetime of about 6×10^7 yr, the lifetime for Fe-bearing dust is 5×10^8 yr (Tielens, 1998 [336]).

7.1.4 Water Masers

There are astronomical observations that provided big puzzles to their interpretation. For example emissions that correspond to surface temperatures of the objects of up to 10^{14} K. These emissions can only be explained by Maser emission.

The acronym MASER denotes Microwave Amplification by Stimulated Emission of Radiation.[4] However, observations have shown that masers emit over a broad portion of the electromagnetic spectrum so the word "microwave" does no longer mean that such phenomena can only be observed in the microwave range.

A LASER is a MASER that works with higher frequency photons in the ultraviolet or visible light spectrum.

Let us briefly discuss the maser effect. When a photon with the appropriate energy is absorbed, electrons are excited to higher levels. The atom absorbs the energy of the photon. Under normal conditions, spontaneous emission occurs: the excited atoms remain only 10^{-8} s in this state and then they spontaneously move to the state before excitation occurred. But there are also states for which the lifetime is much longer e.g. 10^{-3} s, these are called metastable states. By a stimulated emission a photon with a wavelength λ_1 is absorbed by an atom that is already in a metastable state. Two photons of wavelength λ_1 are emitted then. These two emitted photons have exactly the same wavelength, namely λ_1.

Maser radiation can be characterized by:

- Beaming: Imagine an irregularly shaped maser cloud. Since the gain is exponential, any small differences become strongly enhanced. The part of a cloud with a slightly longer path will appear much brighter.
- Rapid variability: this again is a consequence of the exponential gain dependence. A small variation will be therefore exponentially enhanced.
- Line narrowing: the exponential gain amplifies the center of the line shape (which a superposition between Gaussian and Lorentzian profiles). The emission there will be larger, the line appears narrower.
- Saturation: the maser is unsaturated as long as there is no balance between pumping and stimulated emission. When it becomes saturated amplification of radiation depends linearly on the size of population inversion and the path length.
- Temperatures: the brightness is a measure of temperature. An interstellar OH maser at 1665 MHz produces as much radiation as does a black body at a temperature of 10^9 K. Of course at such a high temperature OH molecules would be

[4]The microwaves occur at frequencies between 1 GHz and 300 GHz.

completely dissociated. Some maser brightness even correspond to a temperature of 10^{14} K.

- Polarization: the degree of polarization is usually very high, near to 100%. Such a high degree can be explained by a combination of three effects: (1) Zeeman effect, (2) magnetic beaming of radiation (3) anisotropic pumping.

It should be noted that many of the characteristics of megamaser emission that are found near galactic nuclei are different.

Maser emission has been discovered in different objects such as in molecular clouds, comets, planetary atmospheres, stellar atmospheres (as an example see Fig. 7.4).

The first non terrestrial maser was found in 1965 at a frequency of 1665 MHz (Weaver et al., 1965 [355]). From the intensity of the emitted radiation an extremely high temperature of the object would be derived and therefore, before knowing about the maser mechanism this radiation was a puzzle. Later it became clear that was produced from an OH maser source. In 1969 the first H_2O maser was detected. These molecules can be found in compact sources.

The H_2O maser emission from the source W49 is so strong that in the small bandwidth of only 50 kHz almost a full solar luminosity is emitted and the brightness corresponds to a temperature $>10^{14}$ K (Burke et al., 1970 [51]).

While for a long time the radio line near 22 GHz was the only known water maser transition, in recent years various other water maser lines have been detected at millimeter and submillimeter wavelengths, some of them from the vibrationally excited bending mode. Masers can be emitted by several molecules: The following species have been observed in stimulated emission from astronomical environments: OH, CH, H_22CO, H_2O, NH_3, CH_3OH, SiS, HC_3N, SiO, HCN, H.

Maser detection in the interstellar medium was described by Elitzur, 1987 [118].

Water maser emission at 22 GHz may be a hint for cometary impacts that occur in extrasolar planetary systems. Such radiosignatures were proposed by Cosmovici et al., 2001 [79]. During the impact of comet Shoemaker-Levy 1994 on Jupiter observations made with the 32-m dish Medicina radiotelescope (Bolgna, Italy) revealed the delivery of water and for the first time also a water maser in the solar system. After the comet's nucleus exploded in the upper part of Jupiter's atmosphere, this maser was produced.

Water masers are also found in galaxies and star forming galaxies.

A database of water maser variability over 20 years in a large sample of star-forming regions was done by Felli et al., 2007 [130]. 43 masers in star forming regions were studied at 22 GHz with the above mentioned Medicina radiotelescope. More than 200 sources of water masers are listed in the Arcetri catalogue (first published in 1990, then in 1994 [44]). Very often such water masers coincide with compact IR sources. For young stellar objects it is found that about 25% show both a compact IR source and a water maser. Water masers can occur at the earliest phases of the formation of high luminosity stars and they seem to be connected with molecular outflows (Codella, Felli and Natale, 1996 [72]). Water maser in young stellar objects were catalogued by Furuya et al., 2003 [136] using data from the VLA and the 45-m Nobeyama telescope.

Fig. 7.3 Part of the Very Large Array (VLA). Image credit: Image courtesy of NRAO/AUI

The Very Large Array, VLA, one of the world's most important astronomical radio observatories, consists of 27 radio antennas in a Y-shaped configuration on the Plains of San Agustin fifty miles west of Socorro, New Mexico (Fig. 7.3). Each antenna is 25 meters (82 feet) in diameter. The antennae can be used as an interferometer by combining the data to give the resolution equivalent to an antenna size of 36 km across, with the sensitivity of a dish 130 meters in diameter.

7.2 Water in Starforming Regions

Water is an important molecule in circumstellar material as well as in the interstellar medium. There are known strong far infrared transitions. Therefore it can act as a coolant in shocks and circumstellar outflows. The cm-wavelength water vapor signatures on the other hand indicate star forming regions and cool stars.

Observations of the FIR (far infrared) signatures can only be made from space since the water vapor in the Earth's atmosphere absorbs such signals.

7.2.1 Clouds and Cloud Collapse

About 50% of interstellar gas is concentrated into denser interstellar gas clouds that occupy about 2% of the volume of the Galaxy. The temperature of the gas clouds is about 100 K, the densities range from 1 to 100 atoms per cubic centimeter.

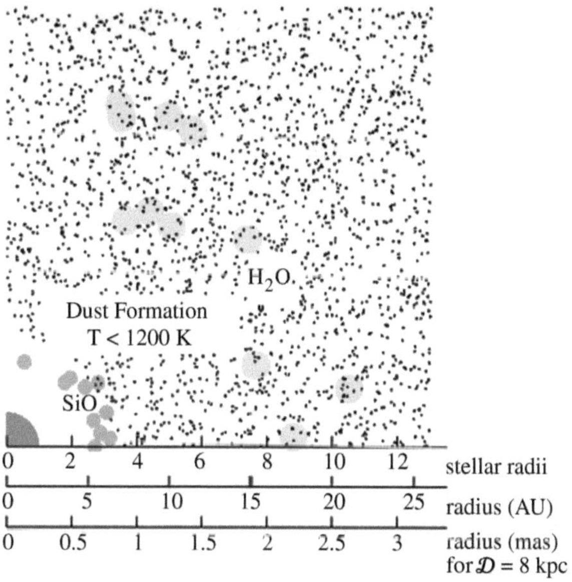

Fig. 7.4 Cartoon representation of the location of SiO and H_2O masers in the inner envelope of a Mira variable. The star is located in *the lower left corner*. The abscissae give from top to bottom the radial distance from the star in stellar radii, astronomical units, and milliarcseconds if observed at the Galactic center distance. SiO masers arise from within a few stellar radii of the central star, while H_2O masers from the vibrational ground state arise from regions somewhat further out in the envelope, where dust formation is occuring. Image credit: Menten, K.M., MPIfR (Sub)millimeter Astronomy Group

In the interstellar space molecules do not survive easily: if the gas is too hot they would collide with other atoms or molecules and break into their constituent atoms. In neutral hydrogen clouds, the temperature may be low enough but another mechanism may destroy them: photons of high energy from neighboring stars. As a consequence, molecules survive only in cold dark interstellar clouds where dust blocks photons. These are the molecular clouds.

Inside such molecular clouds temperatures are only up to 10 K and the densities are from 100 to 1000 molecules per cubic centimeter. The most important molecule is hydrogen, H_2. Other molecules were mentioned already, such as CO and even complex organic compounds are found there. The masses of these molecular clouds are up to 10^7 solar masses and large clouds may be over 1000 light years in size. There exist about 4000 molecular clouds in our galaxy.

How are these clouds affected by their environment? One mechanism is the heating by UV radiation from massive hot stars. Another mechanism are blast waves from supernova explosions. In most molecular clouds self gravity is unimportant. Each part of the cloud is attracted by every other part. But the clouds are not in hydrostatic equilibrium like stars. In most clouds, pressure is much stronger than gravity. Such a pressure causes them to expand, however the expansion is stopped by their surroundings. Some clouds are massive enough so that they start to collapse under their own weight. It is also important that these clouds are cool enough so that the pressure remains low and they really contract.

For a cloud to collapse, the so called Jeans criterion has to be fulfilled: Let us assume a cloud with mass M, radius R, total number of particles N, average particle mass $\overline{m}$, and temperature T. The gravitational potential energy of this cloud is

$$E_{\text{grav}} = -\text{const} \times \frac{GMN\overline{m}}{R} \tag{7.1}$$

The value of the constant depends on the mass distribution inside the cloud but we can take a value 1 for simplicity.

Against gravity, the internal pressure works. The total kinetic energy within the cloud is

$$E_{\text{kin}} = \frac{3}{2}NkT \tag{7.2}$$

From these two equations it becomes evident that a cloud starts to collapse as soon as $|E_{\text{grav}}| > E_{\text{kin}}$ which gives the Jeans mass, M_J and the Jeans density, ρ_J:

$$M_J \sim R\frac{kT}{G\overline{m}} \tag{7.3}$$

$$\rho_J \sim \frac{1}{\pi M^2}\left(\frac{kT}{G\overline{m}}\right)^3 \tag{7.4}$$

Let us do an example: if $M_{\text{cloud}} \sim 2 \times 10^{34}$ kg and $T \sim 30$ K, then the critical Jeans density becomes:

$$\rho_J = 10^{-23}\,\text{kg}\,\text{m}^{-3} \tag{7.5}$$

Thus relatively large massive clouds become gravitationally unstable. The collapse of a cloud is slowed down by the conservation of angular momentum. This causes a flattening of a spherically contracting cloud. Also turbulence and magnetic fields slow the collapse. The smaller the cloud, the stronger becomes the gravity because the gravitational acceleration g depends on

$$g = \frac{GM_{\text{cloud}}}{R^2_{\text{cloud}}} \tag{7.6}$$

and R_{cloud} decreases because of the collapse. From (7.6) we see that when the cloud is $1/4$ as large as it started to be, the force of gravity will become 16 times as strong.

Since molecular clouds are not uniform, some regions in the cloud are denser than others. Therefore, a cloud does not collapse into a single dense object but it starts to fragment into a number of dense cloud cores. These cloud cores are compact with a diameter of only a few light months and from them, the stars finally form. The core begins falling inward faster than the rest of the cloud and the falling material feeds an accretion disk. A protostar is formed at the center. The protostar is heated to several 1000 degrees because gravitational energy of the original molecular cloud is converted into thermal energy. The protostar is also large, therefore it is several thousands of times more luminous than the Sun. However its temperature in its interior is still too low for nuclear reactions to begin.

Such a luminous protostar cannot be observed easily. Since it is cool, most of its radiation is in the infrared. And the protostar is hidden by a dense and dusty molecular cloud. This dust blocks light effectively. However the dust is also heated by the protostar and starts to radiate in the IR. As material is falling on the protostar it adds weight to its outer layers and the material in the interior becomes more and more compressed and therefore hotter. As the protostar shrinks the forces of gravity become greater. If the protostar is massive enough its interior will become heated to several million degrees and nuclear fusion will start. This happens only for stars with at least 0.08 $M_\odot$. Stars with smaller masses will not reach the required 10 million K at their centers in order for nuclear fusion to ignite. These stars are called Brown Dwarfs.

From this short summary about star formation we learn that stars are born from large molecular clouds, they are always born in clusters (also called stellar associations or open clusters—they disintegrate over time). In the Hertzsprung-Russell diagram (H-R diagram) the process of star formation from the collapses of a protostar to the point where it reaches the main sequence where nuclear fusion ignites is also called Hayashi track.

Cloud-collapse and stellar evolution depend on stellar mass. The higher the mass, the faster the collapse and the subsequent evolution. To descend the Hayashi track it takes

- for a 0.1 $M_\odot$ star up to 100 million years,
- for a star with 1 $M_\odot$ about 10 million years,
- for a 10 $M_\odot$ star it takes only 100 000 years,
- for a 100 $M_\odot$ star it takes 10 000 years.

The entire time it took for the Sun to form out from a molecular cloud, fragmentation of the cloud and collapse was about 30 million years.

7.2.2 H_2O *Masers in Star Forming Regions—A Model*

In this subsection we present a proposed model for H_2O masers in a star forming region that was proposed by Elitzur, Hollenbach and Mc Kee, 1989 [119]. As mentioned already, water masers clearly indicate star forming regions. Maser spots are individual clumps. These stream away from the center of activity with velocities up to 200 km/s. Some features reach sizes up to 10^8 km and temperatures between 10^{12} and 10^{13} K. For the pumping of the maser a dissipation of the relative kinetic energies of the shocked and nonshocked gas interstellar shocks provide the energy. In the shocks high temperatures are produced and thus H_2O molecules. It was shown that maser emission could arise in so called C-shocks ($v_s < 40$–50 km/s) or in J-shocks ($v_s \geq 40$–50 km/s). For the column densities: hydrogen about 10^{22} cm^{-2} and H_2O about $\sim 3 \times 10^{19}$ cm^2. The temperature is about 400 K.

How is the H_2O produced? The relevant reactions and the reaction rate coefficients (cm^3 s^{-1}) are given in Table 7.2 (after Elitzur).

Table 7.2 Reactions and reaction rate coefficients for the formation of water in star forming regions

Reaction	Reaction rate coefficient
$H_2 + OH \rightarrow H_2O + H$	$3.6 \times 10^{-11} \exp(-2590/T)$
$H_2 + O \rightarrow OH + H$	$3 \times 10^{-14} \exp(-4480/T)$
$H + OH \rightarrow O + H_2$	$1.4 \times 10^{-14} \exp(-3500/T)$
$H + O_2 \rightarrow OH + O$	$3.7 \times 10^{-10} \exp(-8450/T)$
$H + H_2O \rightarrow OH + H_2$	$1.5 \times 10^{-10} \exp(-10250/T)$

There are several equilibrium equations such as:

$$\frac{\mathrm{O}}{\mathrm{OH}} = 0.47 \frac{\mathrm{H}}{\mathrm{H_2}} \exp(-980/T) \tag{7.7}$$

The rates for production and destruction of H_2O are given as:

$$\frac{\mathrm{H_2O}}{\mathrm{OH}} = 0.24 \frac{\mathrm{H_2}}{\mathrm{H}} \exp(-7660/T) \tag{7.8}$$

The photodissociation of H_2O is given as:

$$\chi = 3.2 \times 10^{-10} \exp(-1.68 A_v)\ \mathrm{s}^{-1} \tag{7.9}$$

χ is the enhancement of the UV flux around 160 nm above the general interstellar field and A_v the visual extinction. For a more detailed discussion see the paper by Elitzur, 1979 [117].

7.2.3 Water Signatures in Protostars

Water signatures in protostars have been observed already 30 years ago. Hydroxyl and Water Masers in Protostars were discussed e.g. by Litvak, 1969 [204]. The first detection of interstellar water molecules was made by Cheung et al., 1969 [67]. They analyzed the 1.35 cm microwave emission in the sources Sgr B 2, Orion and W 49. For numerical calculations, de Jong, 1973 [100] used the abundance of water relative to hydrogen as 10^{-5} that means for 10^5 molecules or atoms of hydrogen there is one H_2O molecule.

As we have pointed out, during the evolution of protostars a disk forms (as an example see Fig. 7.5). Information about such disks from modeling and comparison with observations were discussed by Elmegreen and Morris, 1979 [121]. They modeled protostellar maser sources as orbiting gaseous disks surrounding massive windy stars. According to the model, parts of the disk may be H_2O masers. The high-velocity features seen in several H_2O sources may be due to the material ejected from the inner regions of the disk by pressure from the wind.

Giant planet formation using shallow water simulations of a gaseous protostellar disk was investigated by Zajac, Ingersoll and Dowling, 1991 [371]. They considered vortices in Jupiter's atmosphere. High pressure vortices form spontaneously there and they are stable in an unstable shear flow. They rotate opposite to the sense of

Fig. 7.5 Protoplanetary disks in the Orion nebula. The field of view is only 0.14 light-years across. The Orion Nebula star-birth region is 1 500 light-years away Credit: Hubble Space Telescope

planetary rotation; inside them, there is high pressure and high mass density. They may even grow by merging with smaller vortices that arise spontaneously in the shear flow. An analogue to such a scenario is the physics in a gaseous nebular disk that surrounds a protostar. Vortices can form in the Keplerian shear flow. Thus giant planet formation in a protostellar disk can be simulated.

Codella and Palla, 1995 [71] reported on a search for water masers in protostellar disks. 160 IRAS sources were selected which define high-mass star forming regions. Water maser emission was found in 11 sources of this sample. This could mean that the population of OB-stars was overestimated by 50%.

The water maser emission associated with the infrared centers IRS 1 and IRS 3 of the NGC 2071IR star-forming region was studied by Seth, Greenhill, and Holder, 2002 [305]. They used water masers as tracers for protostellar disks. NGC 2027 is a reflection nebula in the constellation of Orion.

The young stellar object W33A which is deeply hidden protostar was studied in the wavelength range 2.4–25 μm. The spectrum displays deep H_2O ice absorptions centered at 3.0 μm. A value for the H_2O ice column density could be derived: $N(H_2O) = (1.1 \pm 0.3) \times 10^{19}$ cm^{-2} (Gibb et al., 2000 [139]).

A sample of 29 star forming regions was investigated by Beuther et al., 2002 [28]. The 22.2 GHz water maser was investigated at high spatial resolution and the location was compared with mm dust continuum observations, cm continuum and mid infrared sources. Generally H_2O water maser indicate outflows and circumstellar discs. The stellar H_2O maser model is as follows: the excitation occurs by collisions pumping with H_2 molecules within shocks. These shocks are found in outflows and/or accretion. An example of water detected in a star forming region is shown in Fig. 7.8.

7.2.4 T Tauri Stars

T Tauri stars are a group of very young stars, less than 10 million years old and their masses are less than 3 solar masses. These stars have not reached the main sequence on the H-R diagram, so they are still undergoing gravitational contraction representing an intermediate stage between a protostar and a low-mass main sequence star like the Sun.

Because they are very young, T Tauri stars are found only in nebulae or very young clusters, have low-temperature spectra (G to M type) with strong emission lines and broad absorption lines. They are more luminous than main sequence stars of similar spectral type because they are very extended. Besides their location in young clusters and nebulae, another indication for their extreme youth is their high lithium abundance. For normal stars, lithium is rapidly destroyed in stellar interiors. During the last phases of the pre main sequence evolution, when the protostars are still contracting, a lithium burning (pp reactions) occurs.

The Li burning reactions are:

$$^{1}\mathrm{H} + {}^{6}\mathrm{Li} \rightarrow {}^{7}\mathrm{Be} \tag{7.10}$$

$$^{7}\mathrm{Be} \rightarrow {}^{7}\mathrm{Li} \tag{7.11}$$

$$^{1}\mathrm{H} + {}^{7}\mathrm{Li} \rightarrow {}^{8}\mathrm{Be} \tag{7.12}$$

$$^{8}\mathrm{Be} \rightarrow {}^{4}\mathrm{He} + {}^{4}\mathrm{He} \tag{7.13}$$

Protostars are fully convective so Li gets depleted, especially when the star has reached its position on the main sequence.

T Tauri stars often have large accretion disks left over from stellar formation. Their luminosities are not constant but erratic brightness changes are observed. This is explained by violent activity in the stellar atmosphere, or nearby clouds of gas and dust that sometimes obscure the starlight.

T Tauri stars are classified based on spectroscopic characteristics that arise from their disk properties:

- Classic T Tauri stars have extensive disks that result in strong emission lines.
- Weak-lined T Tauri stars are surrounded by a disk that is very weak or no longer in existence. The weak T Tauri stars are interesting for studying early stages of stellar evolution unencumbered by nebulous material. Some of the absent disk matter may have been transformed to planetesimals, from which planets might eventually form. According to one estimate, about 60% of T Tauri stars younger than 3 million years may possess dust disks, compared with only 10% of stars that are 10 million years old. Infant star relatives of T Tauri stars include FU Orionis stars.

The nearest T Tauri stars occur in the Taurus molecular cloud and the ρ Ophiuchus Cloud, both about 460 light-years (140 parsecs) away.

The star AA Tauri belongs to the group of T Tauri stars. It's mass is 0.8 $M_{\odot}$, the luminosity is 0.8 $L_{\odot}$, the surface temperature is about 4000 K and its age is estimated at 2 million years. It is rotating in about 8.5 days which is three times faster

Fig. 7.6 Artist's impression of AA Tauri. Credit: NASA

than the Sun. An example of the appearance of water signatures in a protostellar disk (artist's impression) is given in Fig. 7.6. The disk is around the young star AA Tauri which is located at a distance of 140 pc. The situation there is assumed to be quite similar to that when the Earth formed. Pinte and Menard, 2004 [263] described a model of the accretion disk around AA Tau. The spectral features of AA Tauri are shown in Fig. 7.7.

Hartmann et al., 1998 [157] studied accretion and the evolution of T Tauri disks. The mean accretion rate for a T Tauri star at the age of one million years is about $10^{-8}\ M_{\odot}\ yr^{-1}$ and there is a clear decline of mass accretion rates $\dot{M}$ with increasing age among T Tauri stars.

Water Around T Tauri Stars

H_2O maser emission associated with T Tauri and other regions of star formation was studied by Knapp and Morris, 1976 [186]. The maser source is placed $0.'9$ away from the T Tauri star itself however no obvious object in the optical could have been identified with the H_2O maser.

Shiba et al., 1993 [308] made a near-infrared spectrophotometry for 52 T Tauri stars in the Taurus-Auriga region in the wavelength range from 0.9 to 2.5 micrometers. In more than half of their used sample, wide and shallow dips due to water vapor are superposed at 1.4 and 1.9 micrometers on the spectra. For the central star a spectrum was assumed and after decomposition of the spectra into central star and disk components, the water vapor bands still remain in the disk component therefore the conclusion is that water vapor is located in the disk component of at least 17 T Tauri stars (33%) of the sample. The temperature of the water vapor is about 2000 K which can be estimated from the width of the absorption bands. Certainly such water molecules could exist at different temperatures over the entire disk. In the outer

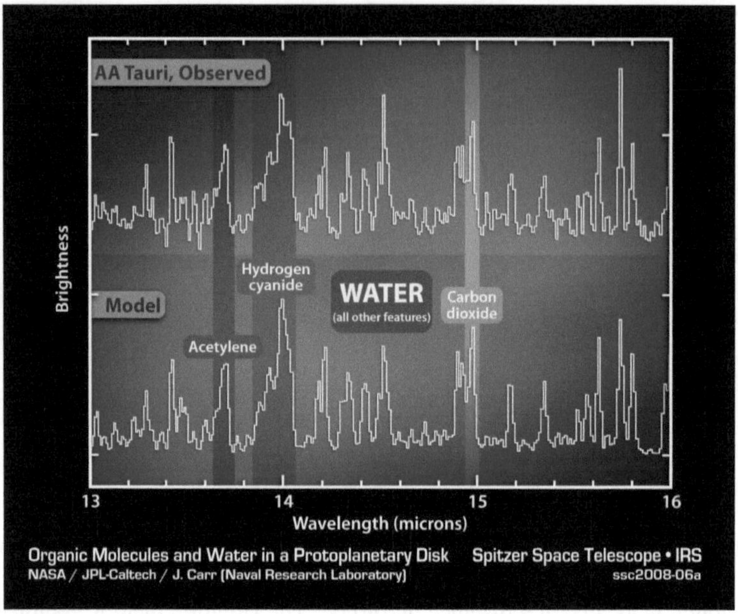

Fig. 7.7 This plot of infrared data shows the signatures of water vapor and simple organic molecules in the disk of gas and dust surrounding a young star AA Tauri. The star is at a distance of 140 pc. Credit: Spitzer Telescope, NASA

regions they would be in solid phase or in dust grains as ice mantles. Where the temperature is higher than the evaporation point (in that case at about 150 K) they are in a gas phase. Bertout et al., 1988 [27] have estimated that such temperatures occur at a distance of 0.03 AU or 6 $R_\odot$ for a central star with 1 $M_\odot$, 3 $R_\odot$ and 4000 K. At even closer distances, the water vapor molecules become dissociated due to the star's UV radiation.

The importance of X-rays on planet-forming molecular environments of T Tauri stars by combining X-ray data (XMM Newton mission) with Herschel far IR data and Spitzer mid infrared data was suggested by Güdel, 2009 [149]. X-rays would also have an influence on the chemistry, e.g. formation of water molecules, in these objects. X-ray heating and ionization of the disk atmosphere around T-Tauri stars were investigated by Glassgold, Meijerink and Najita, 2009 [142]. They show how molecular hydrogen can be formed in the inner disk region, the temperature drops from 1000 to several 100 K. Information on the gas-phase water between 5 and 100 AU around a T-Tauri star can be obtained from Herschel observations (Meijerink et al., 2008 [226]). Herschel (see also last chapter) was launched on 14 May 2009. A 3.5 m Cassegrain telescope performs photometry and spectroscopy in approximately the 55–672 μm range.

7.3 Water Signatures in Spectra of Late Type Stars and the Sun

In astrophysics the term late type stars means stars that are cool (less than 5000 K) and their color is from yellow to orange to red.

7.3.1 Late Type Stars and Water

In this section we give some examples of observations of water signatures in the spectra of late type stars. The stellar evolution will be discussed in Chap. 8. Generally, late type stars are characterized by strong mass loss and in many cases a circumstellar shell forms. At the end of the evolution of stars like the Sun a circumstellar shell is expelled via pulsations, theses ejected layers radiate as planetary nebulae. The strange name planetary nebula comes from the fact that these objects are similar in appearance to giant planets when viewed through small optical telescopes. A typical planetary nebula is roughly one light year across, and consists of extremely rarefied gas, with a density generally around 1000 particles per cm^3.

The presence of water in the atmosphere of cool stars was already predicted by Russell in 1934. His calculations showed that water should be the most abundant molecule besides the H_2 molecule and atomic hydrogen for cool stars at about 2800 K. The first detection of water was made in the spectrum of the well known star Mira. This is a pulsating variable star, an expansion and contraction produces large change in temperature and radius and therefore strong variations in its luminosity are observed. Kuiper (1963) observed Mira in the wavelength range from 1400 nm to 1900 nm.

Searches for 1612-MHz OH and 22 335-MHz H_2O maser emission from symbiotic Mira variables were reported by Seaquist, Ivison, and Hall, 1995 [303]. Because only two cases were found a mechanism for the deficiency was proposed. A powerful wind from the hot companion sweeps much of the volume of the dusty Mira wind away from the hot companion, thus exposing the Mira to UV emission. The UV emission then photodissociates molecules responsible for all known masers.

Barlow et al., 1996 [17], investigated the rich far IR water vapor spectrum of the star W Hya (Fig. 7.9). This object is a cool (2700 K) M-dwarf at a distance of 130 pc. It has a high mass loss (about 10^{-7} $M_\odot\,yr^{-1}$). Water vapor seems to be the major coolant and the H_2O/H_2 abundance is 8×10^{-4} for $r \leq 4.5 \times 10^{14}$ cm and 3×10^{-4} at large radii. Maercker et al., 2008 [210] found high amounts of H_2O in the circumstellar envelope of that star and they speculate that evaporation of icy cometary or planetary bodies might be an effective ongoing mechanism in such systems.

M-star spectra, at wavelengths beyond 1.35 μm are dominated by water vapor. The problem is to distinct spectral signatures coming from the terrestrial water vapor from the stellar signatures. In order to avoid these difficulties, observations with the Infrared Space Observatory can be used. The Infrared Space Observatory, ISO was launched in 1995 by ESA and successfully made 30 000 individual imaging, photometric, spectroscopic and polarimetric observations ranging from objects in our own solar system right out to the most distant extragalactic sources.

Fig. 7.8 The composite picture shows an SWS and an LWS spectrum overlaid on an ISOCAM image of the dark star-forming cloud Rho Ophiuchi. In the high-resolution SWS spectrum the main water feature is the strong emission at wavelengths around 38 microns. The LWS spectrum shows a large number of strong water-vapour emissions at wavelengths from 122 to 183 microns. Credit: ESA

The importance of water vapor lines in the spectra of M stars was addressed by Jones et al., 2002 [174].

Tsuji, 2000 [342] detected water in emission in the spectrum of the M2 supergiant star μ Cephei (M2 Ia) observed by the Short-Wavelength Spectrometer (SWL) on board ISO. The emission first appears in the 6 μm region and then in the 40 μm region (pure rotation lines) despite the rather strong dust emission. The intensity ratios of the emission features are far from those of the optically thin gaseous emission. From the observed emission features an optically thick water sphere of the inner radius about 2 R_* (corresponds to 1300 $R_\odot$), $T = 1500$ K, and a column den-

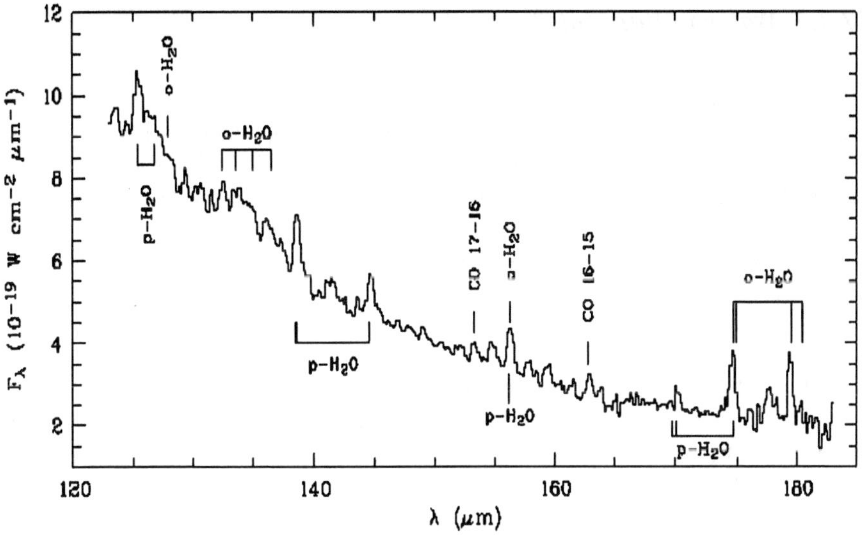

Fig. 7.9 LWS spectrum of W Hya between 122–183 microns, with some of the strongest water vapour lines marked. Credit: ESA

sity of $H_2O = 3 \times 10^{20}$ cm^{-2}. The detection of water in emission provides strong constraints on the nature of water in the early M supergiant star, and especially, its origin in the outer atmosphere is confirmed against other models such as the large convective cell model. The data also indicate that the early M supergiant star is surrounded by a huge optically thick sphere of warm water vapor ($T \sim 10^3$ K).

In asymptotic giant branch stars and red supergiants, masers are mostly observed in stellar winds. Circumstellar water vapor in M-type AGB (asymptotic giant branch) stars using Infrared Space Observatory Long Wavelength Spectrometer data were studied by Maercker et al., 2009 [211]. Such emission lines are important probes of the physics and chemistry in the inner regions of circumstellar envelopes around AGB stars. Water vapor masers in long-period variable stars were analyzed by Winnberg et al., 2008 [362].

A possible relationship between the evolutionary stage of post-AGB stars and planetary nebulae (PNe) and the presence of water masers in their envelopes was investigated by Suárez, Gómez and Morata, 2007 [329]. They used a 70-m antenna in Robledo de Chavela (Spain) to search for the water maser transition at 22.235 GHz. They found two water maser emissions in planetary nebulae (IRAS 17443-2949 and IRAS 18061-2505), a "water fountain" in a post-AGB star (IRAS 16552-3050), and one in a source previously catalogued as a planetary nebula, but whose classification is uncertain (IRAS 17580-3111).

A 22 GHz ortho-water maser in a dusty medium near late-type stars depends on several factors: exchange of energy between dust and gas in the radiation field of a star, radiative cooling by water molecules and pumping of water masers (Babkovskaia, Poutanen, 2006 [14]).

7.3.2 Water in Sunspots?

Sunspots are cool dark regions visible on the solar surface that can extend several tens of thousand km. An example of a big sunspot group and how it moves across the solar disk due to the Sun's rotation is given in Fig. 7.10. Their temperature is about 2000 K less than the surrounding solar surface. They consist of a dark central part, called umbra, surrounded by a filamentary penumbra. Because of their cooler temperatures, molecular lines can form there and such lines have been observed for many years. Measurements of the linewidths can provide information about temperatures and the dynamics. The most prominent features in the mid-IR region are H_2O and SiO lines.

Mallia, Blackwell and Petford, 1970 [213] reported about features attributable to the water vapor molecule in solar spectra. These were found by a balloon borne spectrometer (in order to avoid the disturbing absorption of the Earth's atmosphere) by Woolf et al., 1964 [365] in the following bands: 1.13 μm, 1.4 μm, 1.9 μm and 2.7 μm.

First observations of water signatures in solar spectra were carried out e.g. by Wallace et al., 1995 [352] however they used a Fourier Transform spectrometer with limited resolution. In 1996 Carbon and Goorvitch, [56] reported on water in sunspot umbrae.

Wallace et al., 1996 also prepared a series of four solar spectral atlases. The atlases illustrate the disk center photospheric spectrum from 0.7 to 22 μm and the sunspot umbral spectrum from 1.2 to 21 μm. These spectra were obtained with a Fourier Transform Spectrometer (Mc Math Telescope) at a resolution of 300 000. Many SiO and hot water lines were found.

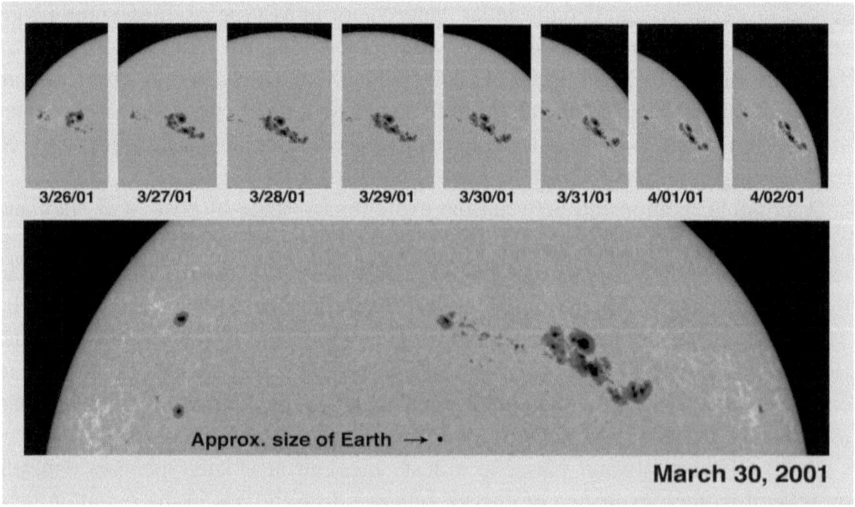

Fig. 7.10 Big sunspot group. Above it is shown how the group passes over the solar disk due to the rotation of the Sun. Credit: SOHO mission, NASA/ESA

Carbon and Goorvitch have computed synthetic sunspot spectra using a H_2O line list and the sunspot umbral models of Maltby et al. (1986) [214].

The infrared spectrum of a sunspot is analyzed in the L-band region 3.1–4.0 μm and 2.02–2.35 μm in the K band. There are 1207 new $H_2^{16}O$ lines assigned in the L band and 508 new lines in the K band (Zobov et al., 2000 [375]).

The abundance of water in sunspots and stars is reviewed shortly by Bernath, 2002 [24]. Sonnabend et al., 2006 [312] made ultra-high-resolution spectroscopic measurements with a resolution of $\sim 10^7$ of water vapor and silicon monoxide in sunspots. Their observations were performed using the Cologne Tunable Heterodyne Infrared Spectrometer (THIS) at the McMath Pierce Solar Observatory. Mid-infrared molecular absorption lines around 10 μm were recorded and resolved in full detail.

7.4 Water in Galaxies

7.4.1 The Milky Way Galaxy

Our Galaxy, the Milky Way, is a barred spiral galaxy and its main components are a disk, a bulge and a halo. The Sun is located about 30 000 light years from the center. The center of the Galaxy is obscured by dust, but with 21 cm measurements of the neutral hydrogen line one can learn about the structure of the Galaxy and its rotation. The diameter of our Galaxy is about 100 000 light years, the thickness of the disk about 1000 light years. The Sun's galactic rotation period is about 220 million years. The number of stars is between 200 and 400 billion. For comparison, if we reduce the solar system (which extends up to the Oort cloud) to a size of 2 mm, the size of the Milky Way would be 100 m.

There are several spiral arms, the Sun belongs to the Orion arm. The spiral arms of our Milky Way contain interstellar matter, diffuse nebulae, and young stars and open star clusters emerging from this matter. On the other hand, the bulge component consists of old stars and contains the globular star clusters mainly in the halo; our galaxy has probably about 200 globulars, of which about 150 are known. These globular clusters are strongly concentrated toward the Galactic Center. An example of a globular cluster is the object M80 (Fig. 7.11). Globular clusters contain very old stars (older than 10 billion years). In the figure one can see some evolved red giant stars. In the interior of these stars oxygen is produced by nuclear fusion and will be released at the end of their evolution. Planetary systems in such cluster are unlikely since the amount of elements heavier than He is extremely small, also water cannot be expected. Cohen and Malkan, 1976 [73] reported about an unsuccessful search for 22-GHz H_2O emission from eight globular clusters.

Fig. 7.11 Globular cluster M80. Credit: NASA, The Hubble Heritage Team, STScI, AURA

7.4.2 Water in the Galaxy?

Besides the distinct sources of water (stars, protostars etc.) surveys were made over larger areas to detect water masers in the galaxy. One of the instruments used for this purpose is the Mopra telescope that belongs to the Australian Telescope National Facility. It is a 22 m single dish telescope at the edge of the Warrumbungle Mountains near Coonabarabran, about 450 km north-west of Sydney. It is used together with other AT antennas (the six 22-m dishes at Narrabri, and the 64-m Parkes dish) as part of the Australian Long Baseline Array. It is also used on its own, for single-dish observations, particularly at millimeter wavelengths. The Mopra is capable of taking spectra while scanning across the sky in either galactic or equatorial coordinates. A relatively large $4'$ by $4'$ field can be mapped in less than one hour.

Walsh et al., 2008, [353] used Mopra Observations to map two one square degree regions along the galactic plane to find water masers. Young stellar objects (YSOs) were studied in the outer Galaxy by detection of water masers (Mochizuki, Hachisuka and Umemoto, 2009 [233]).

7.4.3 Water in Galaxies

What are Galaxies?

Galaxies are huge aggregations of stars, the mass of a galaxy may exceed 10^{11} solar masses. There exist different types of galaxies: spiral galaxies (Fig. 7.12) like

Fig. 7.12 The spiral galaxy M101. Credit: Hubble Space Telescope

our own galaxy, elliptical galaxies, barred spirals and irregular galaxies. This is also called Hubble classification scheme of galaxies. In a spiral galaxy there is a flattened rotating disk. Most stars in the disk of a galaxy move around the center of the galaxy on a nearly circular orbit in the same direction. Stars in elliptical galaxies are moving on complicated orbits; these galaxies take their shapes from the orbits of the stars they contain, the stars move in all directions. Another difference between spiral galaxies and elliptical galaxies is that most spirals contain large amounts of molecular gas and dust. These components are concentrated towards the galactic disk. Elliptical galaxies in contrast contain large amounts of very hot gas which can be observed primarily as X-rays. The stellar density in elliptical galaxies is so high that type I supernovae reheat the gas, it cannot cool off and form a cold cloud that settles to the disk.

For our investigation of water in the universe, star formation and evolution plays a crucial role since elements heavier than boron can only be made during nuclear fusion processes inside stars. In ellipticals, star formation is quite a rare process. The disks of spiral galaxies contain both young and old stars, elliptical galaxies contain a much older population of stars. This can also be seen as a difference in color. The disks of spiral galaxies tend to be blue, because their light is dominated by the luminous young massive hot stars. Elliptical galaxies appear much redder reflecting the fact that they contain an older population of lower-mass stars.

The typical luminosities of galaxies range from 10^6–10^{12} $L_\odot$. Their sizes range from 3000 light years up to several 100 000 light years.

One interesting question is why do spiral arms exist at all? One explanation is that any disturbance in the disk of a spiral galaxy will result into a spiral pattern because of the disk's rotation. Material closer to the center takes less time to complete a revolution around the center than material farther out. Such disturbances can be caused by gravitational interactions with other galaxies. For example our Milky Way galaxy has two satellite galaxies, the Small and the Large Magellanic Cloud.

Galaxies always occur as clusters like shown in Fig. 7.13. The luminous galaxies we see sit within larger, more massive dark matter halos. This can be deduced from

Fig. 7.13 Hubble Illuminates Cluster of Diverse Galaxies (Abell S740), cropped to ESO 325-G004. Note that many other small galaxies are seen on the field. Credit: Hubble Space Telescope

an analysis of their rotation curves.The rotation curves remain flat outside the extend of visible disks. This can only be explained by the presence of matter that does not radiate but that manifests itself by the influence of its gravity. Therefore this is called dark matter. It is even assumed that more than 90% of the mass of a galaxy could be dark matter.

7.4.4 Galaxy Clusters

Our Milky Way galaxy belongs to a small cluster of galaxies which is called the local group (see Fig. 7.14). The are even superclusters of galaxies, like the local supercluster to which our Milky Way and the local group belong to as well as the neighboring Virgo Cluster. There are about 1300 known member galaxies of the Virgo cluster and the total mass is estimated at about 1.2×10^{15} $M_{\odot}$. The radius of the Virgo supercluster is 2.2 Mpc and on the sky it spans about 8 degrees (16 times the diameter of the full moon on the sky) in the constellation of Virgo. Figure 7.14 is an example of a galaxy cluster.

7.4.5 IR-Galaxies

There are galaxies that emit mostly in the infrared, IR. These are called IR-galaxies and they are 10–1000 times brighter in the IR than a typical galaxy (e.g. the Milky

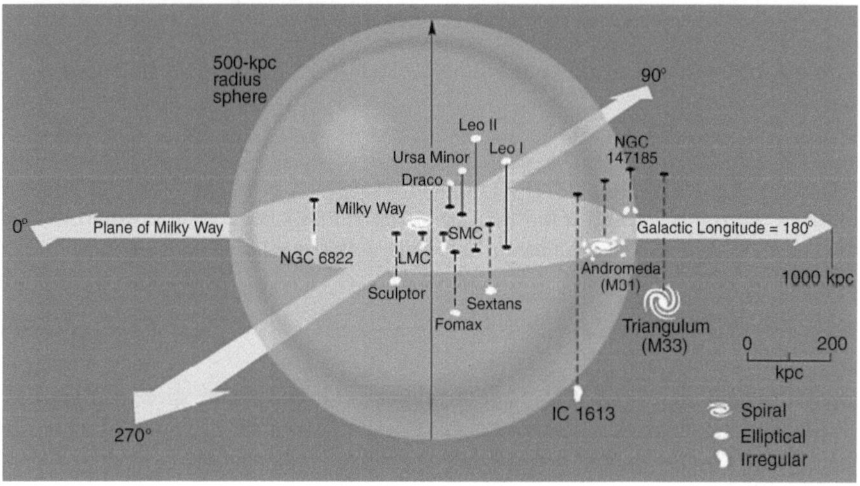

Fig. 7.14 The local group. The most prominent members are our Galaxy and the Andromeda galaxy M31

Fig. 7.15 Color composite image of ultra-luminous infrared galaxy IRAS 19297-0406 made by the Hubble Space Telescope. Credit: HST

Way). An emission in the infrared is almost always associated with the presence of dust. An example is given in Fig. 7.15. Concerning their IR luminosities galaxies can be classified as

- LIRGs: luminous IR-galaxies; about 10^{11} $L_{\odot}$ are emitted in the far IR.
- ULIRGs: the emission in the far IR exceeds 10^{12} $L_{\odot}$.

- HLIRGs: the emission in the far IR exceeds 10^{13} $L_\odot$.

Many of these objects are found at high redshifts z. The redshift is defined as:

$$z = \frac{v}{c} \tag{7.14}$$

where v is the radial velocity of an object that recedes from us. However, this formula would imply that for a redhsift $z > 1$ the velocity of recession would exceed the speed of light c. Thus one has to calculate with the relativistic formula:

$$z = \sqrt{\left(\frac{1 + v/c}{1 - v/c}\right)} - 1 \tag{7.15}$$

Even a value of e.g. $z = 2$ will result in velocities lower than the speed of light.

Many of IR galaxies are starburst galaxies. The star formation rate in these objects is much higher than in normal galaxies. Some contain an active nucleus. There seems to be connections to quasars and the study of ULIRGs means that we deal with galaxy formation at high redshifts. ULIRGs are also embedded in dark matter halos. 10 ULIRGs at a redshift of $z = 2$ were selected and their spectra between 2 and 8 μm were recorded. From these spectra Dust, PAHs (with absorption features at 3.3 and 6.2 μm as well as water ice and hydrocarbons were detected (Sajina et al., 2009 [290]).

7.4.6 Water Masers in Nearby Galaxies

Water masers in a galaxy provide proxies for identifying star forming regions and are therefore of particular interest.

22 GHz water masers were found in four nearby galaxies. The locations of these water masers coincides with ultradense H-II regions. These regions correspond also to bright 24 μm emission areas (see Darling, Brogan and Johnson, 2009 [93] and 2008 [92]).

Using the Expanded Very Large Array, Edmonds et al., 2008 [113] have conducted a search for 22.2 GHz H_2O megamaser emission in the strongly lensed sub mm galaxy, SMM J16359 + 6612 at $z = 2.517$. They did not detect any water maser emission but were able to set an upper limit on the luminosity of any water maser present ($L_{H_2O} \sim 5305$ $L_\odot$).

Using the Green Bank Telescope, Braatz and Gugliucci, 2008 [43] made a survey for water maser emission toward the nuclei of 611 galaxies. They found eight new sources. Their sample consisted of nearby ($v < 5000$ km s^{-1}) and luminous ($M_B < -19.5$) galaxies, some with known nuclear activity but most not previously known to host AGNs (Active Galactic Nuclei).

Three sigma upper limits to the H_2O abundance relative to H_2 ranging from 2×10^{-9} to 1×10^{-8} were given by Wilson et al., 2007 [360] in the galaxies in the galaxies NGC 253, IC 342, M 82, NGC 4258, CenA, and M 51.

7.4.7 *Mega-Masers*

In the nuclear regions of external galaxies, there exist very powerful OH (18 cm) and H_2O (1.35 cm) cosmic masers with line luminosities of 102 104 $L_\odot$, 106 times more luminous than typical Galactic maser sources. These mega-masers are located in high-density molecular gas located in different regions in a galaxy:

- H_2O mega-masers: within parsecs of active galactic nuclei.
- OH mega-masers: within the central 100 pc of nuclear star-burst regions.

H_2O mega-masers are most frequently found in galactic nuclei with Seyfert2 or LINER spectral characteristics, in spiral and some elliptical galaxies. OH mega-masers are found in ultra-luminous IR galaxies (ULIRG) with the warmest IR colors, and importantly, the OH luminosity is observed to increase with the IR luminosity. H_2O mega-maser emission can be mapped at sub-milli-arc-second resolution by Very Long Baseline Interferometry (VLBI). From these observations the distribution and kinematics of molecular gas in distant galactic nuclei at scales below 1 pc can be investigated. From such observations the existence of an accretion disk in Keplerian rotation around a black hole can be proven (e.g. in the object NGC 4258 (see Fig. 7.16), Lo, 2005 [205]).

Fig. 7.16 An artist conception of water maser emission in the accretion disk of the accretion disk and radio jet around the black hole in the heart of the Seyfert galaxy NGC 4258. This accretion disk material lies within a few tenths of a parsec from a supermassive black hole. The disk is warped, rotating differentially according to Kepler's laws. *The inset* at the bottom of the graphic is a radio spectrum (intensity as a function of frequency or velocity) of the water maser emission. The white glints on the disk surface show the locations of regions where maser emission has been detected. Image courtesy of NRAO/AUI and Artist: John Kagaya (Hoshi No Techou)

With the increasing sensitivity of new telescopes and receivers, surveys and high-resolution studies of mega-masers and giga-masers will be very important tracers and high-resolution probes of active galactic nuclei, dust embedded star-bursts in the earliest galaxies and galaxy mergers in the epoch of very active star formation at $z \sim 2$ and beyond. These observations provide also a method for determining the Hubble constant which is important for the expanding universe (see Chap. 8).

Chapter 8
Water—Where Does It Come from?

In this chapter we address the question how the two elements hydrogen and oxygen that form a water molecule were formed during different evolutionary states of the universe and stars.

8.1 The Evolution of the Universe

There are several observational facts that allow us to trace back the evolution of the universe. For more details the reader is referred to the huge amount of textbooks on observational cosmology.

8.1.1 An Expanding Universe

In the 1920s E. Hubble and his coworkers recognized that the spectral lines of galaxies were shifted systematically to the red with respect to the position of laboratory wavelengths. The redshift z of a galaxy can be written as z, v_r is the radial velocity:

$$z = \frac{\lambda_{\text{obs}} - \lambda_{\text{lab}}}{\lambda_{\text{lab}}}, \qquad v_r = cz \tag{8.1}$$

λ_{obs} denotes the observed wavelength, λ_{lab} the rest wavelength, the index lab means the laboratory wavelength.

The systematic redshift of distant galaxies can be interpreted as an expansion of the universe and leads to the famous Hubble law:

$$v = dH_0 \tag{8.2}$$

where d is the distance of a galaxy (in Mpc) and $H_0 \sim 70\ \text{km}\,\text{s}^{-1}\,\text{Mpc}^{-1}$ Hubble's constant at present (indicated by the index 0). This law can be established by measuring the distance of galaxies and their radial velocities. Distance indicators are

A. Hanslmeier, *Water in the Universe,* Astrophysics and Space Science Library 368,
DOI 10.1007/978-90-481-9984-6_8,

Cepheids, Novae, Supernovae and other bright objects. For these objects their true luminosities are known, therefore, the distance can be calculated by comparing their measured brightness with the true luminosity. There exist many other methods for the determination of distances of galaxies and again the reader is referred to the textbooks.

From Hubble's law of expansion we can easily establish an upper limit for the age of the universe, this is called the Hubble time T_H:

$$T_H = \frac{d}{v_r} = \frac{d}{d H_0} = \frac{1}{H_0} \tag{8.3}$$

In the standard cosmological model, the value of the Hubble time is 4.35×10^{17} s = 13.8×10^9 yr.

The future evolution of the universe depends on whether the expansion will continue forever or, due to the presence of enough mass (cosmologists say that the mass density exceeds a critical value), the expansion will be stopped by gravitational attraction and a big crunch will result in the future. The critical density has a value of 8×10^{-27} kg.

Accelerated Expansion

From observations of distant galaxies by the Hubble Space Telescope and other big telescopes evidence for an accelerated expansion was found. This can be explained by a repulsive force and is taken into account by Ω_Λ. A non zero value of Ω_Λ means that the universe is not controlled by gravity alone. In fact Λ was the cosmological constant that was introduced by Einstein in order to keep the solutions of his field equations stable—at the time of propagating his general relativity theory (1916) it was not known that the universe expands.

8.1.2 Radiation from the Early Universe

G. Gamow and R. Alpher predicted a radiation that should be observable distributed uniformly in the universe. When the universe was very young it must have been very small. When gas is compressed, it grows hotter. Therefore, they imagined a hot dense gas in the early universe. The radiation from such a hot gas can be described by Planck's law since it is black body radiation. Due to the expansion of the universe, this radiation would have been red-shifted to longer and longer wavelengths. According to Wien's law, the temperature associated with Planck radiation is inversely proportional to the peak wavelength:

$$T = 2.9[\mu\text{m}]/\lambda_{\text{peak}} \tag{8.4}$$

Shifting the wavelength to longer and longer values is therefore the equivalent of shifting the characteristic temperature of the radiation to lower and lower temperatures. Doubling the scale factor of the universe is equivalent to cutting the temperature of the Planck spectrum in half. The cosmic background radiation, CBR, was detected in 1965 by Penzias and Wilson.

When the Universe Became Transparent

In an ionized hot plasma (e.g. inside the Sun and stars) the free electrons strongly interact with radiation, and its progress gets blocked. Therefore, the hot early universe was not transparent. As the universe expanded, the gas cooled. At an age of about 100 000 years, when the size of the universe was about 10^{-3} of its current size, the temperature dropped to a few 1000 of K. Protons and electrons were able to combine to form hydrogen atoms. This is also called *recombination* of the universe. The hydrogen atoms interact less effective with radiation, the universe became transparent. Since the time of recombination, radiation was able to travel unimpeded throughout the universe. At the time of recombination, the radiation was peaked around 1μm. Now it is peaked around 1 mm (i.e. a factor of 1000).

In 1989 the Cosmic Background Explorer, COBE, was launched. It made precise measurements of the CBR at wavelengths from a few micrometers out to 1 cm (Fig. 8.1). COBE unambiguously proved the CBR to follow a Planck curve at 2.73 K so there is no doubt that we are seeing the residual radiation left behind from the primordial Big Bang. Tiny variations in the CBR result from the formation of structures during the early evolution of the universe (see also Fig. 8.2).

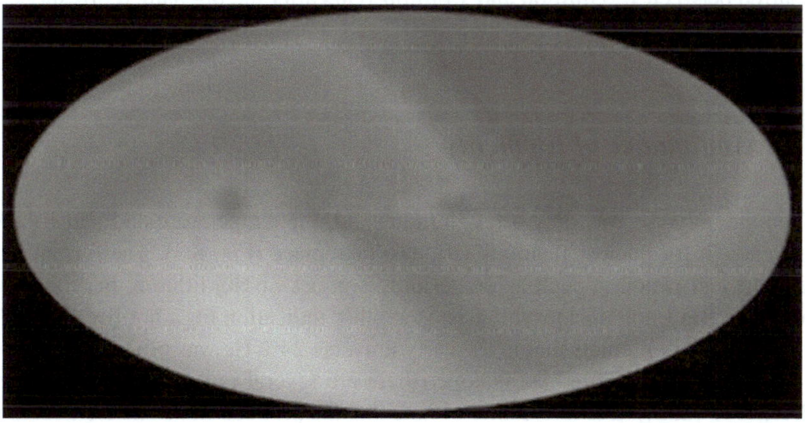

Fig. 8.1 This image of the microwave sky was created from COBE DMR data products. The COBE satellite, launched in 1989, was developed by NASA's Goddard Space Flight Center to measure the diffuse infrared and microwave radiation from the early universe. In this image, the galactic center is in the middle, with the plane of the Milky Way running horizontally across (from Cygnus on the left to Orion on the right). It shows the dipole, a smooth variation between relatively hot and relatively cold areas (upper right to lower left), which is due to the motion of the solar system relative to distant matter in the universe. The signals attributed to this variation are extremely small, only one thousandth the brightness of the sky. Credit: This image of the microwave sky was created from COBE DMR data sets. NASA

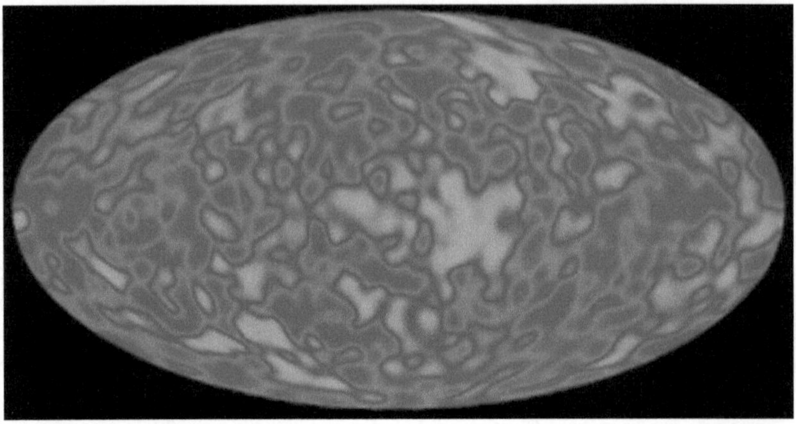

Fig. 8.2 (Color online) The regions which were slightly more dense gravitationally attracted photons and, as the universe expanded, caused them to lose some energy, or heat. The less dense regions are slightly warmer. The map below shows "hot" (*magenta*) and "cold" (*blue*) regions in the radiation detected in the portion of the sky observed. Credit: This image of the microwave sky was created from COBE DMR data sets

Reionization

When the first objects in the universe (probably massive population III stars) started to radiate, the universe went from neutral to become ionized again. This reionization occurred between 150 million and 1 billion years after the big bang (corresponds to $6 < z < 20$).

8.1.3 Abundance of Elements

When the universe was only a few minutes old, the temperature and density were high enough for nuclear fusion reaction to take place (Fig. 8.3). The formation of elements and isotopes, such as deuterium, isotopes of He, lithium, beryllium and boron, the Big Bang nucleosynthesis, is another indication for a hot Big Bang. The Big Bang theory predicts that in the early universe 24% He was produced. This was in the form of the stable ^{4}He isotope (there are several reviews on that topic e.g. Olive, 1999 [253]). It is interesting to note that the production of this element is not strongly dependent on the density of matter in the universe. By density we mean here the value for the today's universe. Between the interval of density of matter from 10^{-29} to 10^{-26} kg m^{-3} the ^{4}He abundance remains very close to the value of 24%.

The other elements and isotopes however sensitively depended on the density of normal matter in the universe. For example the amount of deuterium varies from 10^{-2} to 10^{-10} when the density runs from 10^{-29} to 10^{-26} kg m^{-3}. The ^{3}He abundance varies from 10^{-4} to 10^{-5} for the above mentioned density values.

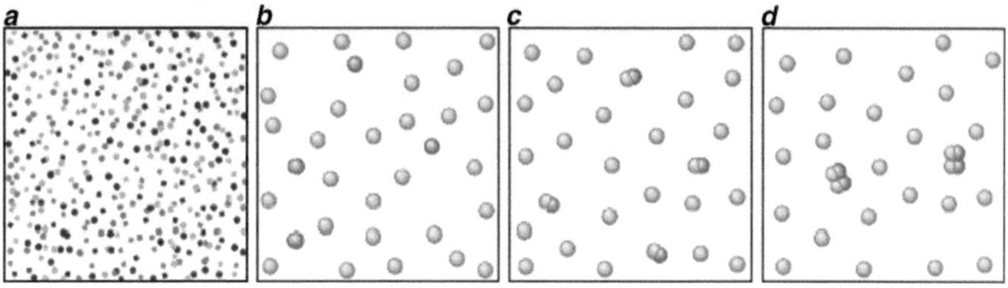

Fig. 8.3 (Color online) The formation of atomic nuclei, started instants after the big bang, as the universe cooled, when the fundamental particles called free quarks (**a**) condensed into protons and neutrons (**b**). Protons (*red*) and neutrons (*blue*) paired off to form deuterons (deuterium nuclei), but because the former outnumber the latter, most of the protons remained alone and became hydrogen nuclei (**c**). Almost all deuterium combined to form helium nuclei (**d**). Therefore there remained only little deuterium to be detected today

All observations are consistent with a density of these elements and isotopes as measured. So the distribution of the isotopes and elements leads to a value of 3×10^{-28} kg m^{-3} for the average density of normal matter in the universe today.

One has to stress again that no elements more massive than boron have been formed in the Big Bang. In this context we also mention dark matter. If dark matter consists of normal matter i.e. of protons and neutrons, then the density of protons and neutrons in the early universe would have been much higher and the resulting abundances of the light elements and isotopes mentioned would have been different from what we measure. A review about dark matter was given e.g. by Bertone, Hooper, and Silk, 2005 [26].

8.1.4 No Water in the Early Universe

We have seen in this section, that there occurred a short phase in the early universe when density and temperature was high enough for the fusion of hydrogen into heavier elements up to lithium and boron. Hydrogen (protons) was present since protons were formed form quark particles. Neutral hydrogen since the recombination age. However, the second constituent of water molecules, oxygen, was not formed during these phases. In the next section we will describe the evolution of stars and see how elements heavier than lithium can be synthesized by thermonuclear fusion reactions inside stars.

8.2 Stellar Evolution

The typical chemical composition of a star is demonstrated by giving the values for the Sun in Table 8.1. In this table the elements hydrogen and oxygen are written in boldface style. The two constituents of a water molecule, hydrogen and oxygen are the most abundant and the third most abundant in solar like stars.

Table 8.1 The chemical composition of the Sun

Element	Percent by number	Percent by mass
Hydrogen	92.5	74.5
Helium	7.4	23.7
Oxygen	0.064	0.82
Carbon	0.039	0.37
Neon	0.012	0.19
Nitrogen	0.008	0.009
Silicon	0.004	0.009
Magnesium	0.003	0.06
Iron	0.003	0.13
Sulfur	0.001	0.04
Total of others	0.001	0.03

Stars can be classified according to their spectra (see Chap. 9). The most important parameter for the evolution of a star is its mass.

In a star like the Sun, the energy is produced in its innermost region, the core. At the center of the Sun, the density of matter is about 150 times the density of water ($1000\ \mathrm{kg\,m^{-3}}$) and the temperature is about 15 million K. Because of their high kinetic energy (which is given by the temperature) and because of the quantum tunneling effect, hydrogen fuses into helium by the so called p-p chain. The individual reactions are:

$$
\begin{aligned}
{}^{1}\mathrm{H} + {}^{1}\mathrm{H} &\rightarrow {}^{2}\mathrm{H} + \mathrm{e}^{+} + \nu, \qquad \mathrm{e}^{+} + \mathrm{e}^{-} \rightarrow \gamma + \gamma \\
{}^{2}\mathrm{H} + {}^{1}\mathrm{H} &\rightarrow {}^{3}\mathrm{He} + \gamma \\
{}^{3}\mathrm{He} + {}^{3}\mathrm{He} &\rightarrow {}^{4}\mathrm{He} + {}^{1}\mathrm{H} + {}^{1}\mathrm{H}
\end{aligned}
$$

${}^{1}\mathrm{H}$ denotes a proton, ν a neutrino, e^{+} a positron, γ a gamma-ray photon. The mass of the four separate hydrogen atoms is 1.007 times greater than the mass of a single helium atom. Therefore, when hydrogen fuses to helium, 0.7% of the mass of the hydrogen is converted to energy by Einstein's famous formula:

$$E = mc^2 \tag{8.5}$$

To illustrate the enormous amount of energy that is released let us give a comparison. Fusing just one single gram of hydrogen into helium releases about 6×10^{11} J of energy.[1] In the Sun, about 600 million tons of hydrogen is converted to helium every second. That means that about 4 million tons of matter is converted to energy.

[1] With that energy the water in 10 average swimming pools can be boiled.

8.2.1 Red Giants

The lifetime of a star can be estimated by the ratio of the amount of fuel divided by the rate fuel is used.

$$\tau_{\mathrm{MS}} = (1.0 \times 10^{10}) \times \frac{M\ (\mathrm{M}_\odot)}{L\ (\mathrm{L}_\odot)}\ \text{years} \tag{8.6}$$

The index MS denotes lifetime on the main sequence.[2] This formula yields $\tau_{\mathrm{MS}} = 1 \times 10^{10}$ yr for stars with one solar mass. For the very hot and bright massive O5 stars for example the lifetime is only 8×10^5 years. For the faint less massive M0 stars for example the expected lifetime is 7.9×10^{10} years which is a factor of five of the age of the universe.

After the consumption of the hydrogen in their cores, a hydrogen burning shell develops and stars evolve toward a red giant. It takes 200 million years for a star like the Sun to move from the main sequence to the top of the red giant branch (see Fig. 9.10). The first 100 million years the Sun will be a subgiant with a luminosity of about 10 $\mathrm{L}_\odot$. During he second half (i.e. the second part of the 200 million years mentioned above), the luminosity of the Sun evolves from 10 $\mathrm{L}_\odot$ to about 1000 $\mathrm{L}_\odot$.

From the point of view of nuclear fusion, these two stages can be summarized as:

- Main sequence star: H burning core, non burning H envelope;
- Red giant star: non burning degenerate He core, H burning shell, non burning H envelope.

Such an object is called a red giant. The core of a red giant is stabilized by the pressure of degenerate electrons, where the electrons are packed so densely that quantum mechanics determine their pressure. As the star evolves up the red giant branch, its helium core grows smaller and more massive and also gets hotter. One reason for this heating of the core is that gravitational energy gets released when it shrinks. When the temperature reaches about 10^8 K the collisions between He nuclei in the core become energetic enough to overcome the electric repulsion and He burning starts, also called triple alpha process. First two ^{4}He nuclei fuse to form a beryllium nucleus, ^{8}Be. This nucleus is extremely unstable (lifetime only 10^{-12} s), however if it collides with another ^{4}He nucleus a stable carbon ^{12}C nucleus is formed. Therefore, from three ^{4}He nuclei, one carbon nucleus results.

After the main sequence phase, the next phase in stellar evolution is the ignition of helium, also called helium flash (see Fig. 8.4). The degenerate electrons are very good conductors; any imbalance in temperature is quickly evened out. Therefore, the degenerate core of a red giant star is at almost the same temperature everywhere. When the helium burning begins at the center, the energy is quickly released over the entire core. Within a few minutes therefore, the entire core is burning helium into carbon by the triple alpha process. In a normal gas you would expect that due by the increase of temperature that occurs by the He burning, this gas would

[2] In the Hertzsprung Russell diagram temperature of stars are plotted against their luminosity, most stars are on a diagonal called main sequence.

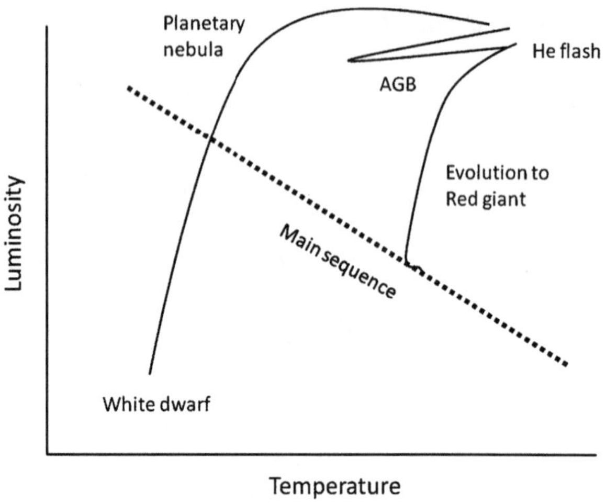

Fig. 8.4 Sketch of the evolution of low massive stars in the Hertzsprung-Russell diagram. On the abscissa T is plotted, on the ordinate the luminosity

expand because of the increased pressure. As a consequence of such an expansion, the temperature, density and pressure in the core would decrease again.

The equation of state for normal gas is:

$$P = nkT, \quad P \sim \rho T \tag{8.7}$$

The pressure depends on both, temperature T and number density n or density ρ. The non relativistic degenerate gas is governed by the followingequation of state:

$$P = K\rho^{5/3} \tag{8.8}$$

and for the relativistic case, the equation of state is:

$$P \sim \rho^{4/3} \tag{8.9}$$

The thermal runaway that takes place during the helium flash is dramatic. Helium burning begins at a temperature of about 100 million K. An explosion starts because as the temperature gets higher, the helium burning reaction rates get higher. Within seconds, the thermal pressure increases to the point that is no longer smaller than the degeneracy pressure, now the core explodes. This explosion cannot be seen since it takes place inside the star. The energy that is released lifts the overlying stellar layers. The core is no longer degenerate, and a new equilibrium has to be found. So one important effect of the helium flash is to remove degeneracy from the core. The core becomes much larger, the acceleration of gravity is small because

$$g = GM/r^2 \tag{8.10}$$

A larger core means smaller values of g and weaker gravity means less weight pushing down on the core, lower pressure. Therefore nuclear reactions slow down, and the star becomes less luminous.

An overview on the structure and evolution of low mass stars was given e.g. by Catelan, 2007 [63].

8.2.2 The Asymptotic Giant Branch

During the next 100 000 years the situation is as follows: in the non degenerate core, helium burns to carbon, the luminosity of the star is now 1/100 as it was at the time of the helium flash. The star shrinks, its surface temperature increases. In the H-R diagram (Fig. 9.10) the stars move to the left of the red giant branch. In this horizontal branch, stars burn helium in the core to carbon and hydrogen to He in a shell. Differences in their chemical composition affect where these stars fall on the horizontal branch.

The strong dependence on small differences in chemical composition and other parameters like for example magnetic field strengths becomes more important toward the last stages of stellar evolution.

Whereas the structure of a horizontal branch star (Fig. 8.4) is much like the structure of a main sequence star, there are two major differences: the energy is generated in two regions (core: He burning, shell: hydrogen burning). Because of the He burning in the core, the carbon ash starts to accumulate. When the horizontal branch star has burned all of the helium at its core, gravity once again starts to win and the core is crushed by the weight of the layers above it. Finally, the electrons become degenerate again. The carbon core is now electron degenerate. The star ascends the asymptotic branch again closely parallel to the path it followed as a red giant. It gets closer and closer to the red giant branch and becomes more luminous. This phase of evolution is therefore called asymptotic giant branch, AGB. An AGB star burns He and H in concentric shells that lie above a degenerate carbon core.

Nucleosynthesis in AGB stars and its relevance for galactic enrichment and planetary system formation was reviewed by Busso, Gallino, and Wasserburg, 1999 [52]. They stressed the importance of neutron capture by a process like $^{12}C(\alpha, n)^{16}O$. This means that a carbon isotope ^{13}C reacts with an α particle and the result is an oxygen ^{16}O plus a neutron, n.

8.2.3 A Carbon Flash?

Now one would expect that a carbon flash will happen. However this is not the case. The star expels its outer layers. When the Sun becomes an AGB star, its outer layer will engulf the terrestrial planets—the Earth would move in the atmosphere of the Sun.[3] Because of $g = GM/r^2$, the acceleration due to gravity at the surface of a star with a mass of one solar mass and a radius of 100 $R_\odot$ is only 1/10 000 as strong as the gravity at the surface of the Sun. Such an object is quite unstable. The instability of such stars of such stars is enhanced by the strong dependence of the triple alpha process (conversion of three He into carbon) on temperature. Episodes of rapid energy release may occur and the star looses up to 30% of its initial mass.

---

[3]This will happen in about 5 billion years from now.

8.2.4 Post AGB Evolution

As we have seen, a star loses mass during its AGB evolution. By losing several % of its mass, the weight of the layers is reduced, the outer layers of the star expand. The star becomes less massive and larger. After ejection of its outer layers, all that is left of the low-mass star itself is a tiny very hot electron degenerate carbon core. The remaining hydrogen and helium rapidly burn to carbon and more and more of the mass of the star ends up in the carbon core. Within about 30 000 years the star moves from right to left across the H-R diagram. The surface temperature may reach up to 100 000 K. According to Wien's law

$$\lambda_{\text{peak}} = \frac{2\,900}{100\,000} = 0.029\ \mu\text{m} \tag{8.11}$$

this radiation is mainly in the UV. This radiation heats and ionizes[4] the expanding shell of gas. A planetary nebula forms. Examples of planetary nebulae are given in Figs. 8.5 and 8.6. The name for such object is completely misleading: they are just the remnants of the outer layers of a star ejected into space. From the complex appearances of these nebulae one may deduce that there were stages in the stellar evolution, when mass loss was much faster and times, when mass loss occurred primarily from the star's equator or its poles. Because of the expansion of the gas, such planetary nebulae are visible for about 50 000 years before the gas ejected by the star disperses so far that the nebula becomes too faint to be seen. For a review on planetary nebuale see e.g. Peimbert, 1990 [259].

8.2.5 Elements Heavier than He

The above mentioned processes and stages of stellar evolution are important in the context of enriching the interstellar matter with elements heavier than He. The presence of such metals[5] in the interstellar matter is a consequence of convection which leads to a mixing of elements. As we have pointed out, in the core He burning produces carbon. By convection, material from the interior (carbon rich) is transported upward and mixed with material in the outer parts of the star. For the present Sun, convection goes down to a depth of only 200 000 km which is about $1/3\ R_\odot$. The steep temperature gradient of AGB stars cause convection to penetrate into deeper and deeper layers of such a star. Therefore, many AGB stars are known as carbon stars because they show an overabundance of carbon and other products of nuclear burning. Planetary nebulae often show an overabundance of elements such as carbon, nitrogen, oxygen which are by-products of nuclear burning.

The gas disperses and the interstellar matter becomes enriched with heavier elements.

[4] Ionized H clouds, H II regions.

[5] Astrophysicists call all elements heavier than He metals.

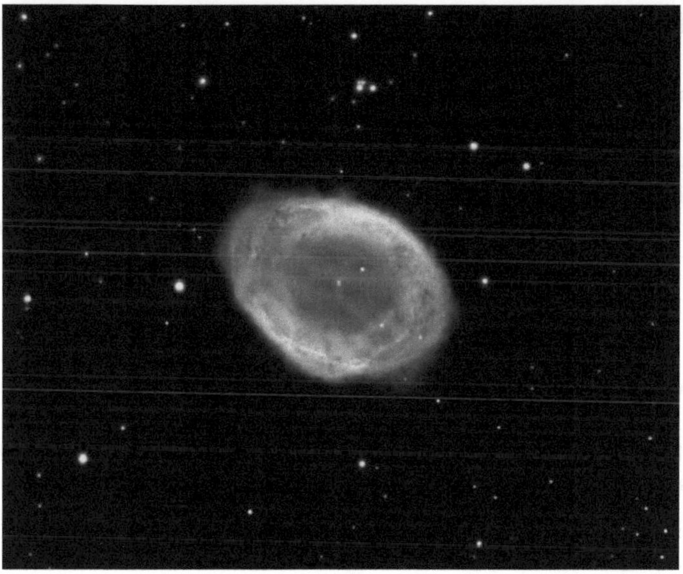

Fig. 8.5 The ring nebula, a very famous planetary nebula. In the center, the white dwarf star is seen

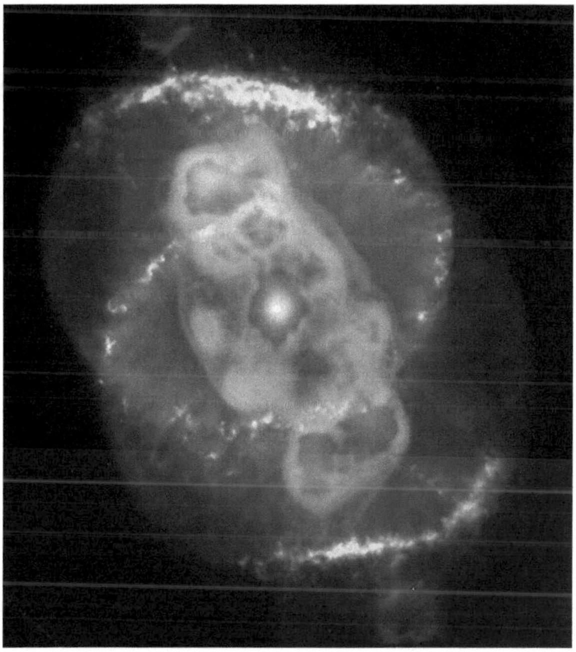

Fig. 8.6 The cats eye planetary nebula. X-ray/optical composite image of NGC 6543, the Cat's Eye Nebula (X-ray: NASA/UIUC/Y. Chu et al., Optical: NASA/HST)

8.2.6 The Ultimate Fate of a Low Massive Star: White Dwarfs

Within 50 000 years a post AGB star leaves behind a cinder of non burning carbon with a mass of about 0.6 $M_{\odot}$. It shrinks to the size of the Earth. Then it again becomes fully degenerate and cannot shrink further.

At this stage we speak of a white dwarf. Such objects radiate and cool, therefore, their temperature remains very high for several ten million years. Their luminosity however is only 1/10 000 that of a main sequence star because of their small size. This will be the ultimate fate of our Sun.

No Water from Low Massive Stars

The contribution to the amount of water observed in the universe from low massive stars is small for two reasons:

- Almost no oxygen is produced during their different phase of thermonuclear burning. Only elements up to carbon are synthesized.
- Their evolution is rather slow, lasting at least several billion years. Thus their overall contribution to the heavier elements is small.

The existence of water in the universe is therefore mainly based on the different nuclear burning phases that occur in the more massive stars.

8.3 Massive Stars

8.3.1 Main Sequence Evolution of Massive Stars

What happens to stars that are more massive? The central temperature is much higher and additional nuclear reactions become possible. One example of such reaction is the CNO cycle. The main reactions are listed below.

$$
\begin{aligned}
{}^{12}\mathrm{C} + {}^{1}\mathrm{H} &\rightarrow {}^{13}\mathrm{N} \\
{}^{13}\mathrm{N} &\rightarrow {}^{13}\mathrm{C} + \mathrm{e}^{+} + \nu \\
{}^{13}\mathrm{C} + {}^{1}\mathrm{H} &\rightarrow {}^{14}\mathrm{N} \\
{}^{1}\mathrm{H} + {}^{14}\mathrm{N} &\rightarrow {}^{15}\mathrm{O} \\
{}^{15}\mathrm{O} &\rightarrow {}^{15}\mathrm{N} + \mathrm{e}^{+} + \nu \\
{}^{15}\mathrm{N} + {}^{1}\mathrm{H} &\rightarrow {}^{12}\mathrm{C} + {}^{4}\mathrm{He}
\end{aligned}
$$

In principle, this reaction is the most important one in stars with more than 1.5 solar mass. There is a further difference between low massive and massive stars concerning their central core. In high massive stars there is a large temperature gradient because of the greater temperature dependence of the CNO cycle, the core becomes

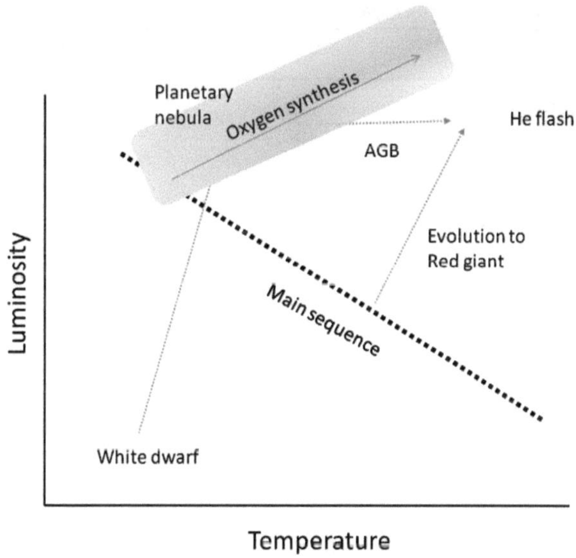

Fig. 8.7 The region where oxygen is synthesized is marked in the HR-diagram. Also the evolution paths of a low massive and a massive star are represented schematically

convectively unstable. Therefore, the He ash is spread uniformly throughout the core. As a high massive star leaves the main sequence, a non degenerate core develops. Long before the core becomes electron degenerate, the temperature reaches 10^8 K which is needed for He burning. The star undergoes a smooth transition from hydrogen burning to He burning. As a high massive star leaves the main sequence, it grows in size while its surface temperature falls. therefore it moves off to the right of the H-R diagram.

When the high massive star exhausts its He in the core, the core starts to collapse until it reaches a temperature of 8×10^8 K. Then carbon begins to burn and more massive elements like sodium or magnesium are formed. Now the star consists of a carbon burning core, a helium burning shell and still farther away from the center a hydrogen burning shell. When carbon is exhausted, Ne burning starts and the oxygen burning (see also Fig. 8.7). Finally, the structure of a massive star becomes as follows (from the surface to the core):

- non burning envelope,
- H burning to He,
- He burning to C,
- C burning to Na, Ne, Mg,
- Ne burning to O, Mg,
- O burning to S, Si,
- S, Si burning to Fe,
- degenerate Fe core.

As a star undergoes post-main sequence evolution, it may make several passes through the instability strip in the H-R diagram. There the star becomes variable, i.e. a Cepheid variable which alternately blows up and shrinks again causing an observable brightness variation.

Table 8.2 Burning stages in massive stars

Core burning	9 $M_\odot$	25 $M_\odot$	Temperature
H burning	20 Myr	7 Myr	$3–10 \times 10^7$ K
He burning	2 Myr	700 000 yr	$1–7.5 \times 10^8$ K
C burning	380 yr	600 yr	$0.8–1.4 \times 10^9$ K
Ne burning	1.1 yr	1 yr	$1.4–1.7 \times 10^9$ K
O burning	8 months	6 months	$1.8–2.8 \times 10^9$ K
Si burning	4 days	1 day	$2.8–4 \times 10^9$ K

Table 8.3 Stages of thermonuclear energy generation in stars

Process	Fuel	Products	T	Min. mass
H burning	H	He	$1–3 \times 10^7$ K	0.1
He burning	He	C, **O**	2×10^8	1
C burning	C	**O**, Ne, Na, Mg	8×10^8	1.4
Ne burning	Ne	**O**, Mg	1.5×10^9	5
O burning	O	Mg, S	2×10^9	10
Si burning	Mg, S	up to Fe	3×10^9	20

The different burning stages depend on the mass of the stars. In Table 8.2 some examples are given. It is seen that oxygen burning lasts only several months.

Iron is the most massive element formed by fusion. With iron, the chain of nuclear fusion stops. Energy is released by nuclear fusion only if the binding energy of the products is greater than the binding energy of the reactants. As we see from Table 8.2 the stages of burning after helium are progressively shorter-lived. Carbon burning is only able to support the star for less than 1000 years for example. There is one important point: the luminosity of a Si burning star is only slightly greater than it was when the star was burning helium. However the neutrinos cool the star effectively (in fact at the stage of Si burning the energy that is emitted in the form of neutrinos is 200 million times the energy of electromagnetic radiation). The most important stages of thermonuclear reaction rates are summarized in Table 8.3. The production of oxygen is highlighted.

8.3.2 Supernova Explosion

The final evolution of massive stars (see e.g. the review given by Woosley, Heger, and Weaver, 2002 [366]) becomes more violent than that for low massive stars. A cosmic catastrophe occurs. The gas of the Fe core becomes electron degenerate. As soon as it exceeds about the mass of 1.4 solar masses (this is called Chandrasekhar mass), i.e. the core exceeds the size of the Earth, gravity is too strong to

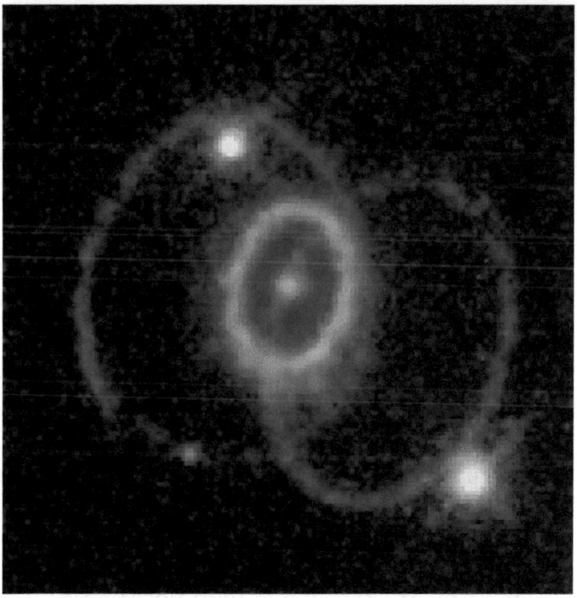

Fig. 8.8 Supernova 1987A. The age of the exploded star was about 20 Myr. Credit: NASA

be compensated by electron degeneracy. Extremely high temperatures, 10^{10} K and densities occur, the radiation is so energetic that thermal gamma rays are produced with break the iron nuclei apart into helium nuclei. This process is called photo-disintegration. The electrons are squeezed into atomic nuclei where they combine with protons to produce neutron rich isotopes in the core of the star. These processes consume up energy and at the same time neutrinos continue to pour out the star. All these events take place during about one second. The collapse of the core occurs at velocities up to 70 000 km/s. Due to the collapse, the density of matter reaches the density of an atomic nucleus, the strong force becomes repulsive. Half of the collapsing core slows its inward fall, half slams into the innermost part of the star and bounces, sending a tremendous shock wave back out throughout the star. Neutrinos are produced in huge amounts, over the next second almost 1/5 of the stellar mass is converted to neutrinos that pour outward through the star's mass. Some of these neutrinos are trapped adding energy to the outward expanding shock wave and within a few hours this shockwave reaches the surface of a star. Now we can observe a so called Type II supernova. The stellar surface is heated up to 500 000 K and the material moves at velocities of up to 30 000 km/s.

The kinetic energy of the ejected material is about 10^{47} J. The amount of energy carried away from the supernova explosion by neutrinos is 100 times larger. In the 20th century a massive star in the Large Magellanic Cloud, a companion galaxy to the Milky Way, exploded, supernova 1987A (Fig. 8.8). Neutrino detectors measured a burst of neutrinos passing through Earth which confirmed the model described above.

For our investigations about the chemical elements in the universe, especially hydrogen and oxygen, supernovae are extremely interesting since they eject newly formed massive elements into space. The interstellar medium becomes enriched with massive elements by mass loss of low and high massive stars. Type I supernovae (they are formed when a companion star accretes mass from its primary and when the mass of this companion star exceeds the Chandrasekhar limit) and type II supernovae are important. The production of heavier elements therefore is done by:

- Low massive stars: elements up to carbon and oxygen are produced.
- Massive stars: elements up to iron are produced.
- Elements heavier than iron cannot be formed by nuclear burning. During a supernova explosion many free neutrons are produced. These can be captured by atomic nuclei and later they decay to protons.

As long as the core left behind the supernova explosion is less than about three solar masses, the collapse will be stopped by the pressure of degenerate neutrons. Thus a neutron star is formed which is a very compact object with a radius of only 10 km. Pulsars are rapidly rotating neutron stars. A famous example of a supernova remnant is the Crab nebula. The shell of this nebula expands at velocities of up to 1500 km/s. The filaments contain anomalously high abundances of helium and other more massive chemical elements. The Crab Pulsar rotates 30 times a second (see Fig. 8.9). The Crab Pulsar (PSR B0531+21) is a relatively young neutron star. The star is the central star in the Crab Nebula, a remnant of the supernova SN 1054 and it was observed e.g. by ancient Chinese astronomers.

Supernova remnants sometimes show water emission (shock excited, e.g. in the supernova remnant 3 C 391, Reach and Rho, 1998 [277]). Woodall and Gray, 2007 [364] investigated 18 known supernova remnants for evidence of the 22.235-GHz water maser spectral line, using the 20-m telescope at the Onsala observatory in Sweden. Their results were negative, which was consistent with their theoretical models. Only in unusual environments can 22-GHz water masers be detected around supernova remnants. The Very Large Array was also used to search for water masers in the three supernova remnants W28, W44, and IC 443. Only one candidate maser was detected, in W44. Follow-up observations using the VLA at 1612 MHz demonstrate that the water maser in W44 is likely associated with a late-type star and not a shock in the supernova remnant (Claussen et al., 1999 [70]).

8.3.3 Stellar Populations

A stellar population is a group of stars that resemble each other in spatial distribution, chemical composition (metallicity) and age. In astrophysics all elements heavier than iron are called metals. Therefore, metallicity denotes the content of elements heavier than helium in a star. When we say e.g. a nebula is rich in oxygen or nitrogen it does not mean that it is rich in these elements from the point of view of chemistry we know on Earth.

The metallicity of an astronomical object may provide a clue to its age. During the Big Bang practically no metals were produced. Therefore older stars have a

Fig. 8.9 (Color online) A composite image of the Crab Nebula showing the X-ray (*blue*), and optical (*red*) images superimposed. The size of the X-ray image is smaller because the higher energy X-ray emitting electrons radiate away their energy more quickly than the lower energy optically emitting electrons as they move. Credit: NASA

lower metallicity than younger stars that are formed out of interstellar matter that was enriched by metals.

Population III stars exist only in theory. It is assumed that they consisted only of the primordial chemical elements H and He. They were extremely massive and therefore they evolved rapidly and ended in a supernova explosion. Rollinde et al., 2009 [279] investigated the influence of population III stars on the chemical evolution of the universe. Population III stars with masses between 30 and 40 $M_\odot$ could explain a sharp decrease of the number of low-mass stars at very low iron abundance, which is not reproduced in models with only PopII/I stars.

The oldest observed stars are population II stars. They have very low metallicities. Stars of population I are the youngest stars with a relatively high metallicity. The metallicity of the Sun is approximately 1.6% by mass.

The distribution of different stellar populations throughout the Galaxy is illustrated in Fig. 8.10.

This distribution can be summarized as:

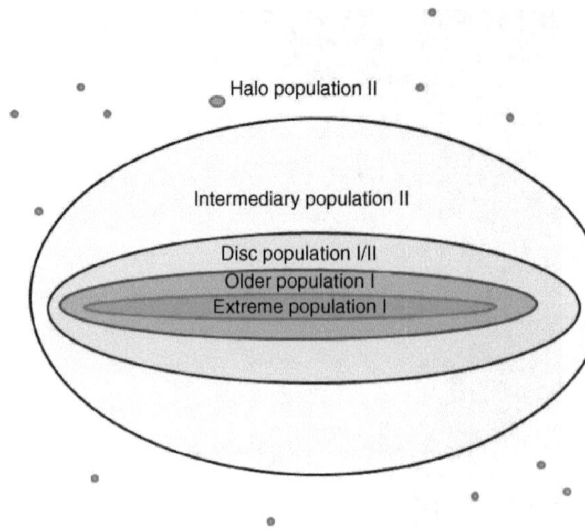

Fig. 8.10 Distribution of stellar populations over the Galaxy

- Thin disk: here we find the youngest stars; O and B stars, T Tauri stars. The age is from 0 to 5 billion years; they are close to the disk, the extend of their distribution from the disk is up to 250 pc. The abundance of metals relative to the Sun is 11.
- Thick disk: these stars have ages between 5 to 14 billion years; they are G and K stars and planetary nebulae; the distribution extends to 700 pc from the disk and their metal content relative to the Sun is $1/3$.
- Halo: most of the stars are in globular clusters, RR Lyrae star s (a group of variable stars like the Cepheids). They are about 11 billion years old and their distribution extends up to 2500 pc. Their metallicity relative to the Sun is $1/30$.

Therefore we can expect to find signs of water only for stars of the younger population I. This is also important for the detection of extrasolar planets which might be very rare for the old population II stars. The transition from population III stars which might have been the first stars in the universe and injected metals to the interstellar medium to population II and population I stars is described by Maio et al., 2010 [212]. The population III regime lasts for a very short period during the first stages of star formation ($\sim 10^7$ yr). Today, we cannot observe population III stars since they all evolved rapidly due to their high masses (these are estimated to be even greater than 100 $M_{\odot}$).

Chapter 9
Appendix

9.1 How to Detect Water

In this chapter we start with a short overview about the spectral signatures of water that allow us to detect water on solar and extrasolar system objects.

9.1.1 Transparency of the Earth's Atmosphere

Our atmosphere is not transparent to all wavelengths of the electromagnetic spectrum. Absorption in the Earth's atmosphere occurs by different molecules.

- O_2, O_3 (ozone) absorb almost all wavelengths $\lambda < 300$ nm.
- H_2O absorbs many wavelengths $\lambda > 700$ nm.

There are only two "windows" where the atmosphere becomes transparent to electromagnetic radiation. The so called optical window comprises the whole range of visible light (400–700 nm) and is in the range from 300 nm (this corresponds to the UVC band) to around 1100 nm. As it is shown below, there exist several small windows in the IR.

The radio window ranges from about 1 cm to about 11 m wavelengths.

A schematic representation of absorption in the IR and Radio domain is given in Fig. 9.1. This figure illustrates the relatively small domain of visible radiation.

Water and NH_3 absorb at wavelengths <8 μm, gases like CO_2 or CH_4 absorb at longer wavelengths. Bands in the IR are listed in Table 9.1.

Whereas the detection of water on extrasolar system objects (stars, interstellar medium, galaxies) is based on spectroscopic investigations, for objects in the solar system also in situ measurements to detect water can be made.

A. Hanslmeier, *Water in the Universe,* Astrophysics and Space Science Library 368, DOI 10.1007/978-90-481-9984-6_9,

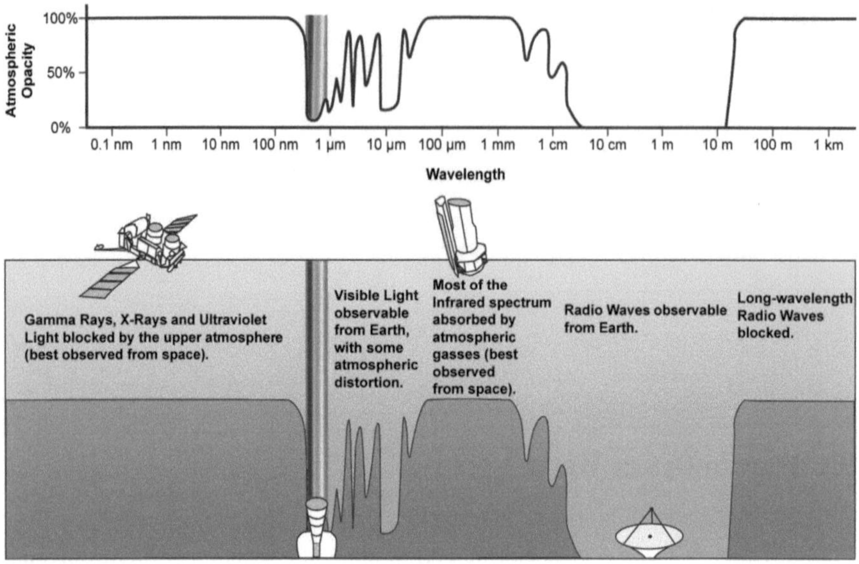

Fig. 9.1 Transparency of the Earth's atmosphere. Credit: isc.astro.cornell.edu/spoon/crashcourse/

Table 9.1 Bands in the IR used in astronomy

Wavelength	Band	Instruments
0.65 to 1.0	R, I	All major optical telescopes, OT
1.25	J	Most major OT, and most infrared telescopes IT
1.65	H	OT, IT
2.2	K	OT, IT
3.45	L	OT, IT
4.7	M	OT, IT
10	N	OT, IT
20	Q	OT, IT
450	submm	Submillimeter telescopes

9.1.2 In Situ Measurements

Water on solar system planets like Mars can be detected by measuring gamma-ray and neutron fluxes. For Mars, a gamma ray spectrometer, GRS, was used to determine the abundance and distribution of elements like hydrogen on its surface. Gamma rays are emitted from the surface of a planet when it is exposed to cosmic rays. When such atoms are hit by the high energy cosmic ray particles they emit neutrons which scatter and collide with other atoms. These atoms get excited then and during the transition to their normal states, a gamma ray photon is emitted. The energy of the emitted rays depends on the elements present in the surface material.

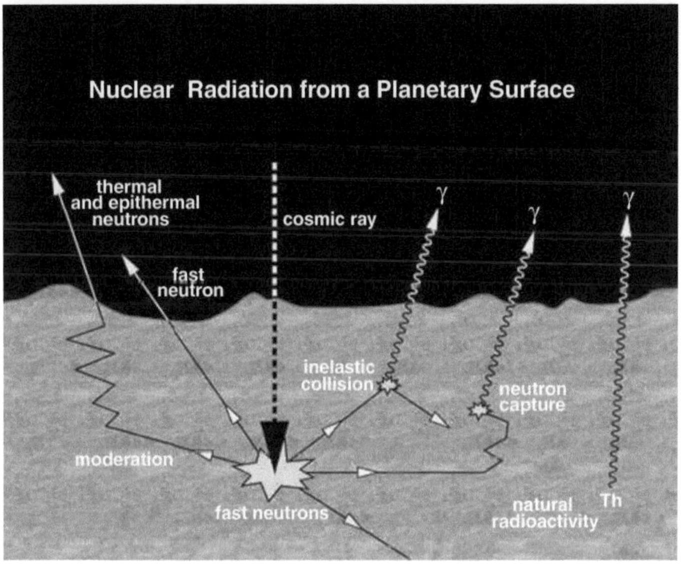

Fig. 9.2 Gamma ray emission process due to cosmic ray excitation of surface atoms. Courtesy: NASA

From the intensity of the emitted radiation, the abundance of such surface elements can be derived. This process is illustrated in Fig. 9.2.

For example hydrogen emits a strong signal at 2223 keV due to prompt gamma emission following neutron capture on hydrogen (Boynton et al., 2002 [42]).

9.1.3 Spectroscopic Signatures

Now we discuss how water molecules can be detected in the spectra of astrophysical objects.

Molecular Spectra

In order to understand the spectroscopic signatures of water, we shortly introduce some concepts how molecular spectra can be understood and interpreted. In the Born-Oppenheimer Approximation it is assumed that the motions of the electrons and the nuclei of a molecule can be separated. If $\mathbf{r}_i$ denotes the positions of the electrons i and $\mathbf{R}_j$ the positions of the nuclei j, then the wavefunctions are given by

$$\Psi_{\text{molecule}}(\mathbf{r}_i, \mathbf{R}_j) = \Psi_{\text{electrons}}(\mathbf{r}_i, \mathbf{R}_j \Psi_{\text{nuclei}}(\mathbf{R}_j)) \tag{9.1}$$

Because of the higher mass, the nuclear motion is much smaller than the electron motion, the nuclei can be considered fixed. So the electronic wavefunction depends

Fig. 9.3 Molecular spectra

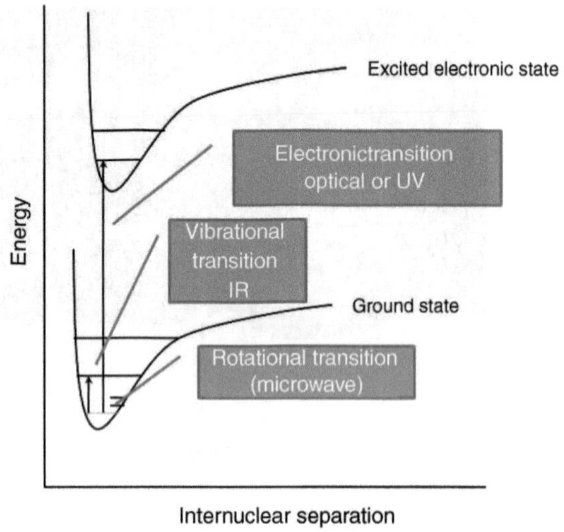

only on the nuclear positions $\mathbf{R}_j$ but not on the nuclear velocities. The nuclear motion which could be rotation or vibration sees something like a smeared out potential from the much faster moving electrons.

Let us consider a diatomic molecule HCl. Electronic states can be represented by plots of potential energy as a function of internuclear distance. Electronic transitions occur very rapidly, so the internuclear distance will remain nearly constant during the transition. Therefore, as it is shown in Fig. 9.3, the electronic transitions are nearly vertical. There exist three types of transitions:

1. Electronic transitions in the optical or UV. These occur between excited states and the ground state.
2. Vibrational transitions: They occur between different vibrational levels in the same state and can be observed in the IR.
3. Rotational transitions: they occur mostly between rotational levels of the same vibrational state. These transitions can be observed in the microwave region.

If a molecule has a dipole moment, as it is the case for water, incident electromagnetic waves excite the rotational levels of such molecules by exerting a torque on the molecule. The spectral signatures for such rotational transitions are usually observed in the microwave region (see also Fig. 9.4).

If x, y, z are the principal axes of rotation and I_x, I_y, I_z the corresponding moments of inertia, the rotational kinetic energy is:

$$E_{\text{rot}} = \frac{1}{2}\left(I_x^2 + I_y^2 + I_z^2\right) \tag{9.2}$$

Or in terms of angular momenta:

$$E_{\text{rot}} = \frac{L_x^2}{2I_x} + \frac{L_y^2}{2I_y} + \frac{L_z^2}{2I_z} \tag{9.3}$$

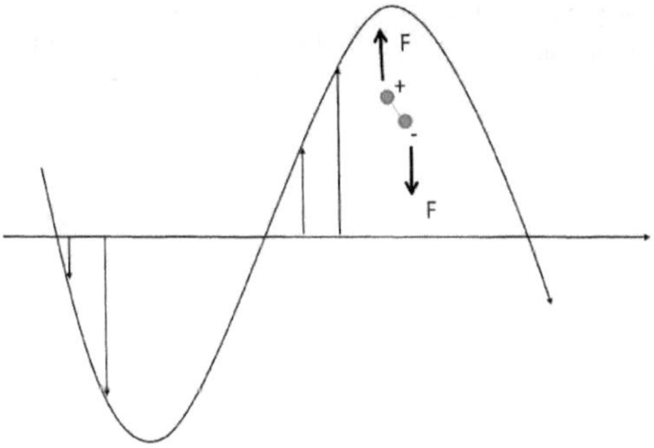

Fig. 9.4 Interaction of an electromagnetic wave with a dipole molecule

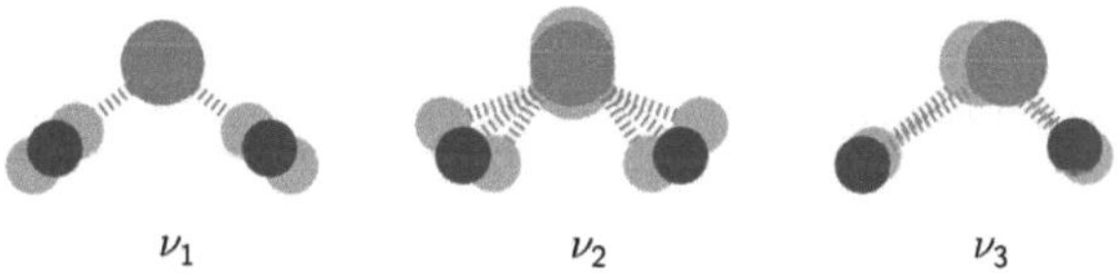

Fig. 9.5 Main vibrations occurring in water

In order to calculate the Hamiltonian of a freely rotating molecule, the angular momenta have to be replaced by the corresponding quantum mechanical operators, e.g. for the z-component:

$$L_z = -i\hbar\frac{\partial}{\partial\Phi} \tag{9.4}$$

The above mentioned considerations are valid only for diatomic molecules. Water, H_2O, consists of three atoms. Therefore the situation becomes more complicated. This is illustrated by Fig. 9.5.

There occurs:

- symmetric stretch, ν_1,
- asymmetric stretch, ν_3,
- bending, ν_2.

Because of the low weight of the hydrogen atoms, the vibrations have large amplitudes. The moment of inertia of water molecules is low. Thus there is a large manifold of combined rotational-vibrational spectra.

Some examples of these transitions are given in Table 9.2.

Generally, the environment plays an essential role on the spectra. Let us consider an increase in temperature. The raise in temperature causes a weakening of the hydrogen bonds whereas the covalent O–H bonds strengthen. As a result, the molecular stretch vibrations shift to higher frequencies whereas the intramolecular vibrations shift to lower frequencies as well as the molecular bend frequencies.

In Table 9.3 the IR vibrational spectrum of liquid water is given.

In liquid water and ice librations occur (restricted rotations). These librations are due to the hydrogen bonding. Therefore the IR and Raman spectra are more complex than those of vapor. To give an example: for liquid water at 0°C the minor L_1 band

Table 9.2 Main vibrations of water isotopologues

Gas	ν_1 cm^{-1}	ν_2 cm^{-1}	ν_3 cm^{-1}
$H_2^{16}O$	3657.05	1594.75	3755.93
$H_2^{17}O$	3653.15	1591.32	3748.32
$H_2^{18}O$	3649.69	1588.26	3741.57
$HD^{16}O$	2723.68	1403.48	3707.47
$D_2^{16}O$	2669.40	1178.38	2787.92
$T_2^{16}O$	2233.9	995.37	2366.61

Table 9.3 Main vibrations of water isotopologues

Wavelength	cm^{-1}	Assignment
0.2 mm	50	intermolecular bend
55 μm	183.4	intermolecular stretch
25 μm	395.5	L_1 librations
15 μm	686.3	L_2 librations
6.08 μm	1645	ν_2 bend
4.65 μm	2150	$\nu_2 + L_2$
3.05 μm	3277	ν_1 symmetric stretch
2.87 μm	3490	ν_3 asymmetric stretch
1900 nm	5260	$a\nu_1 + \nu_2 + b\nu_3; a + b = 1$
1470 nm	6800	$a\nu_1 + b\nu_3; a + b = 2$
1200 nm	8330	$a\nu_1 + \nu_2 + b\nu_3; a + b = 2$
970 nm	10310	$a\nu_1 + b\nu_3; a + b = 3$
836 nm	11960	$a\nu_1 + \nu_2 + b\nu_3; a + b = 3$
739 nm	13530	$a\nu_1 + b\nu_3; a + b = 4$
660 nm	15150	$a\nu_1 + \nu_2 + b\nu_3; a + b = 4$
604 nm	16500	$a\nu_1 + b\nu_3; a + b = 5$
514 nm	19460	$a\nu_1 + b\nu_3; a + b = 6$

is at 395.5 cm^{-1}, the major L_2 band at 686.3 cm^{-1}. The absorbance of L_1 increases with increasing temperature, the L_2 absorbance decreases but broadens. Ice has a sharper major band at 819 cm^{-1} at -10°C.

The detection of water molecules outside the solar system became an interesting task for radio astronomy since in 1963 the Hydroxyl, OH was detected. This OH can be formed through a photolysis of water.

The lowest possible pure rotational transition of water vapor is the 5_{23}–6_{16} line. This line can be found at a frequency of 22.2351 GHz or 1.35 cm.

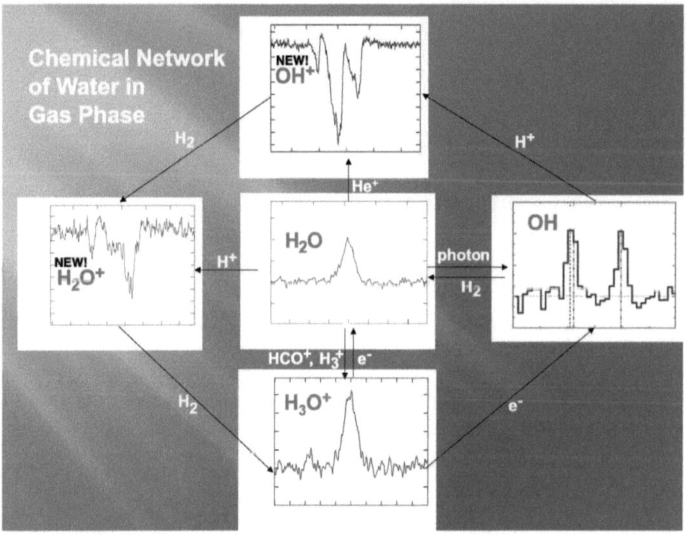

Fig. 9.6 The complex reactions between molecules to form water. Courtesy: Arnold Benz

Formation of Water Molecules

In Fig. 9.6 the complex of chemical reactions that can take place for the formation of water is given. There are three reactions:

- Denotating gas reaction:

$$2H_2 + O_2 \rightarrow 2H_2O \tag{9.5}$$

- Ion reaction involving OH^+, H_2O^+ and H_3O^+.
- Catalytic reaction on the surface of interstellar dust.

For example atomic beams of oxygen and deuterium are aimed at a porous water ice substrate (H_2O) held at 10 K and the production of HDO and D_2O was observed (Dulieu et al., 2010 [110]).

The Water Hole and Interstellar Communication

Between the radio frequencies 1420 and 1640 MHz there lies the so-called water hole (Fig. 9.7). Interstellar hydrogen can be detected at 1420 MHz. This corresponds to a wavelength of 21 cm. At 1660 MHz (18 cm) there appears the signature of OH. Both H and OH can be regarded as dissociation products of water. B. Oliver proposed in the 1970s (see e.g. Oliver, 1979 [254]) that the spectral region between these two lines could be used for interstellar communication since it is a very quiet region where no other spectral signatures are found. In this range of frequencies the background galactic noise is also low.

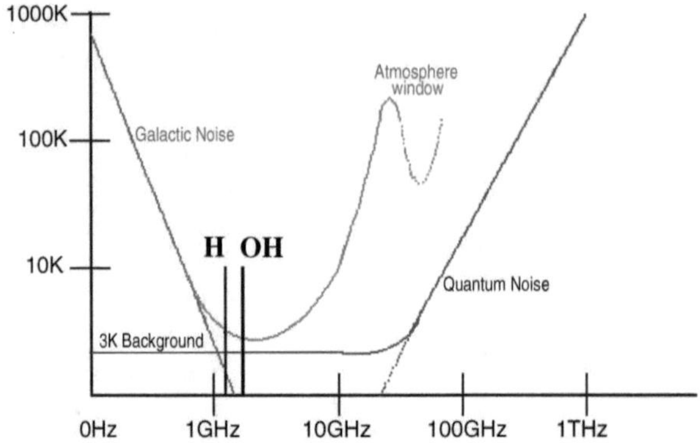

Fig. 9.7 The water hole. Within this Microwave Window, photons travel relatively unimpeded through the interstellar medium, the background noise is at minimum. Credit: The SETI League, Inc

Stull, 1975 [328] stressed the necessity to protect this frequency range and pleaded for an international legal protection of the "Water Hole" frequency band for a search of extraterrestrial intelligent life. The SETI association and other similar programs therefore look for spectral signatures in this water hole. Any signatures observed there could be only of artificial origin. Davoust, 1991 [98] gave an overview of the cosmic waterhole. In Davoust, 1988 [97] the "silence" in the water hole is discussed.

An indirect or 'passive' search for extraterrestrial intelligence (SETI) in 25 globular clusters observed at the water and hydroxyl lines was made by Cohen, Malkan and Dickey, 1980 [74] with negative results.

9.2 Satellite Missions

As it was pointed out already, water can be detected by IR measurements and most of these observations have to be made from satellites because of the absorption of the Earth's atmosphere.

9.2.1 Water Detection with SWAS

The Submillimetre Wave Astronomy Satellite, SWAS, was launched on December 5 1998 and is operated by the Harvard-Smithsonian Center for Astrophysics as one of NASA's Small Explorer Projects (SMEX). It was designed to study the composition of the interstellar gas, mainly water, molecular oxygen and carbon molecules.

The following features can be observed:

- Water, H_2O at 556.936 GHz,
- Molecular oxygen, O_2 at 487.249 GHz,

Table 9.4 IR satellites

Name	Year	Entrance window size cm
IRAS	1983	57
ISO	1995–1998	60
Spitzer	2003–2008 + warm	85
Akari	2006–2007 + warm	67
Herschel	2009–	3.5 m

- Neutral carbon CI at 492.161 GHz,
- Isotopic carbon monoxide ^{13}CO, at 550.927 GHz,
- Isotopic water $H_2^{18}O$ at 548.676 GHz.

A short review of the main results from SWAS:

- In regions of space where the temperature is only 30 K, water was found to be far less plentiful than expected.
- In hot regions, mainly star forming regions, the water concentration seems to be as much as ten thousand times larger than expected before.
- The amount of oxygen is much lower than expected.
- Water observed in the atmospheres of Jupiter and Saturn is thought to come from cometary bombardment of these gas giants.
- The distribution of water in the Martian atmosphere matches a profile of constant, 100% saturation from 10 to 45 km altitude.

9.2.2 IR Satellites

In Table 9.4 a list of IR/sub mm satellites is given. The size of the entrance window as well as the duration of the mission and the range of the detector are also listed. The term warm means, that the satellite will be used after the cooling (e.g. by He) has terminated.

Herschel (formerly known as FIRST, Far InfraRed and Sub-millimetre Telescope) was successfully launched on 14 May 2009 by ESA. The telescope was named after Sir W. Herschel who proved the existence of IR radiation. Herschel carries a 3.5 meter diameter passively cooled telescope. Herschel (Fig. 9.8, see also Doyle, Pilbratt, and Tauber, 2009, [111]) is the only space facility dedicated to the submillimeter and far infrared part of the spectrum (55 to 672 μm range). The main tasks of Herschel are:

- Study the formation of galaxies in the early universe and their subsequent evolution.
- Investigate the creation of stars and their interaction with the interstellar medium.
- Observe the chemical composition of the atmospheres and surfaces of comets, planets and satellites.
- Examine the molecular chemistry of the universe.

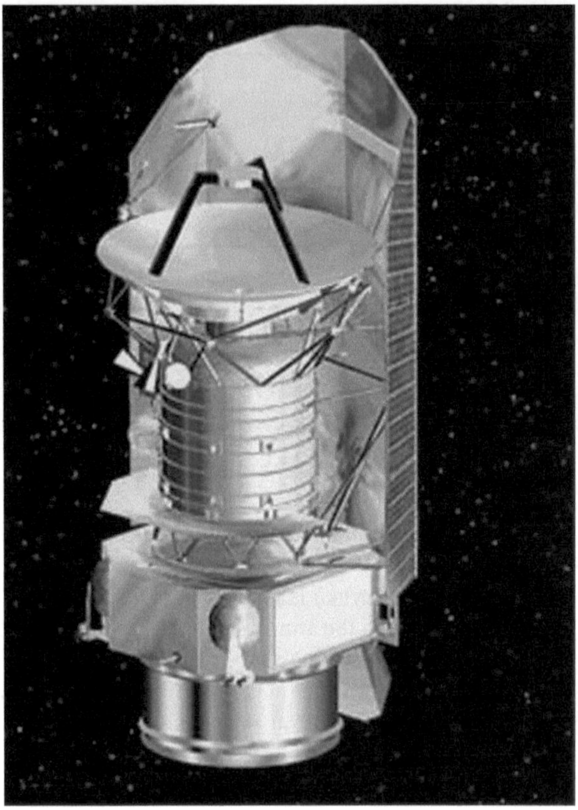

Fig. 9.8 The Herschel Space Telescope. Credit: ESA

9.2.3 Future Astronomical Telescopes

The detection and observation of extrasolar planets will be one of the key projects in astrophysics in the next decades. Several instruments will be available to observe the faint signatures of planets and to study by spectroscopy their atmosphere and to detect water therein.

One of these projects is the European Extremely Large Telescope (E-ELT) (see Fig. 9.9). Earth sized planets will be discovered and the atmospheres of giant planets can be studied with these new fascinating facilities. The detection of the planets will be made by precise velocity measurements and larger planets can be directly observed. The telescope will have a diameter of 42 m and be operated by ESO, the European Southern Observatory.

The James Webb Space Telescope (JWST) is a large, infrared-optimized space telescope. JWST will have a large mirror, 6.5 meters (21.3 feet) in diameter and a sunshield the size of a tennis court.[1] The telescope will be placed at one of the

[1]This shielding prevents from heating which causes IR noise.

Fig. 9.9 The E-ELT that will be operated by ESO. Credit: ESO

Lagrangian points 1.5 million km away from Earth. This is a point where the gravitational forces of the Sun and Earth are equal. The birth of stars and protoplanetary systems and planetary systems and the origins of life will be two important research fields with that instrument. The JWST will be able to measure spectra of disks around young stars that contain icy particles. The Mid-Infrared Instrument (MIRI) is an imager/spectrograph that covers the wavelength range of 5 to 27 micrometers, with a possible spectrographic coverage up to 29 micrometers (compare with Table 9.3). The Near Infrared Camera (NIRCam), provided by the University of Arizona is an imager with a large field of view and high angular resolution. The NIRCam covers a wavelength range of 0.6 to 5 micrometers and has ten mercury-cadmium-telluride (HgCdTe) detector arrays. These are analogous to CCDs found in ordinary digital cameras.

9.3 Some Astrophysical Concepts

9.3.1 *Apparent Magnitude*

In astrophysics, the brightness of a star is given in magnitudes. The apparent brightness of a star depends on its true brightness and its distance. The definition is as follows: two stars with a difference of 1 magnitude, written as 1^{m}, differ by a factor of 2.512 in their intensities:

$$m_1 - m_2 = -2.5 \log \frac{I_1}{I_2} \tag{9.6}$$

Table 9.5 The spectral classes of stars

Type	Temperature	Mass	Radius	Luminosity
O5	44 500	60	17.8	800 000
B0	30 000	18	9.3	5 500
A0	9250	2.9	2.5	54
F0	7200	1.6	1.35	6.5
G0	6000	1.05	1.05	1.5
G2 (Sun)	5860	1.0	1.0	1.0
K0	5250	0.79	0.85	0.4
M0	3850	0.51	0.63	0.08

where m_1, m_2 are the two measured apparent magnitudes, I_1, I_2 the intensities. Thus a star of magnitude 6 is $2.512^5 = 100$ times fainter than a star of magnitude 1. Stars with magnitude 6 are the faintest stars, that one can see with the naked eye under favorable conditions (no light, clear dark skies). The absolute magnitude M of a star is its apparent magnitude at a defined distance of 10 pc (32.6 light years). It can be shown, that the difference between apparent and absolute magnitude (also called distance modulus) is related to the distance of the object as follows:

$$m - M = 5\log d - 5, \quad [d] = pc \tag{9.7}$$

where d denotes the distance in parsec (pc). So one has to know M and measure m in order to find the distance.

For example, if one detects a supernova in a galaxy, and assuming, that its absolute magnitude is -18^M, and the measured apparent magnitude is about $+14^m$, then the distance of the host galaxy is:

$$m - M = 14 - (-18) = 5\log d - 5 \tag{9.8}$$

The luminosity of a star with radius R and temperature T depends on its surface and the energy output that is given by the Stefan Boltzmann law:

$$L = 4\pi R^2 \sigma T^4 \tag{9.9}$$

where $\sigma = 5.67 \times 10^{-8}\,\mathrm{J\,s^{-1}\,m^{-2}\,K^{-4}}$.

9.3.2 Spectral Classes

The spectra of stars can be classified according to their temperature in the following sequence: O-B-A-F-G-K-M. The classes are subdivided into 0–9. This sequence is a temperature sequence, the O stars being the hottest. In the spectra, O stars exhibit few lines, cooler stars many lines, even molecular bands because of their lower temperature. Basically, there is no difference in chemical composition between the different spectral types.

The main properties are listed in Table 9.5. The values for mass and luminosity are given in solar units.

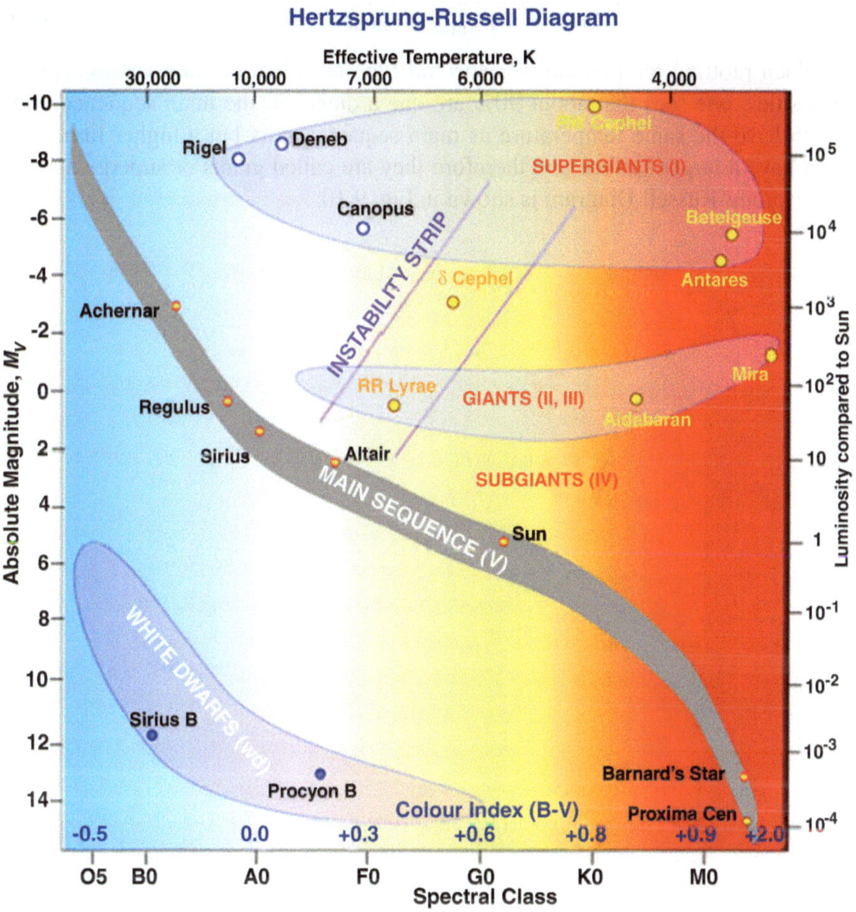

Fig. 9.10 The Hertzsprung-Russell Diagram. In the abscissa, the quantity plotted B-V means color which is a measure for temperature. The temperature decrease from left (about 50 000 K) to right (about 3000 K). The hydrogen-burning dwarf stars (about 90% of all stars) like the Sun are found in a band running from top-left to bottom-right called the Main Sequence. Giant stars form their own clump on the upper-right side of the diagram. Above them lie the much rarer bright giants and supergiants. At the lower-left is the band of white dwarfs. Credit: Image source: http://outreach.atnf.csiro.au/education/senior/astrophysics/stellarevolution_hrintro.html

9.3.3 *The Hertzsprung-Russell Diagram, HRD*

The color of a star is a measure for its temperature, blue stars are hotter than red stars. This is a consequence of Wien's law that states that the peak of emission of a black body shifts toward shorter wavelengths when the temperature is higher:

$$T\lambda_{\max} = \text{const} \tag{9.10}$$

When plotting temperature (or spectral type or color) of stars versus their luminosities, one sees that about 90% are one a diagonal, the main sequence. Stars which have the same temperature as main sequence stars but a higher luminosity must have a larger surface and therefore they are called giants or supergiants. The Hertzsprung-Russell Diagram is shown in Fig. 9.10.

References

1. Abe, Y., Numaguti, A., Komatsu, G., Kobayashi, Y.: Four climate regimes on a land planet with wet surface: effects of obliquity change and implications for ancient Mars. Icarus **178**, 27–39 (2005)
2. Abramov, O., Spencer, J.R.: Endogenic heat from Enceladus' south polar fractures: new observations, and models of conductive surface heating. Icarus **199**, 189–196 (2009)
3. Adams, F.C., Laughlin, G.P.: The frozen Earth. In: Bulletin of the American Astronomical Society, vol. 31, p. 849 (1999)
4. A'Hearn, M.F.: Deep impact and the origin and evolution of cometary nuclei. Space Sci. Rev. **138**, 237–246 (2008)
5. Allen, M.R., Ingram, W.J.: Constraints on future changes in climate and the hydrologic cycle. Nature **419**, 224–232 (2002)
6. Allison, M.L., Clifford, S.M.: Ice-covered volcanic water flows on Ganymede. In: Lunar and Planetary Institute Conference Abstracts, vol. 15, pp. 5–6 (1984)
7. Allison, M.L., Clifford, S.M.: Ice-covered water volcanism on Ganymede. J. Geophys. Res. **92**, 7865–7876 (1987)
8. Alvarez, L.W., Alvarez, W., Asaro, F., Michel, H.V.: Extraterrestrial cause for the Cretaceous Tertiary extinction. Science **208**, 1095 (1980)
9. Andrews-Hanna, J.C., Zuber, M.T., Arvidson, R.E., Wiseman, S.M.: Early Mars hydrology: meridiani playa deposits and the sedimentary record of Arabia Terra. J. Geophys. Res. (Planets) **115**, 6002 (2010)
10. Arnold, J.R.: Ice in the lunar polar regions. J. Geophys. Res. **84**, 5659–5668 (1979)
11. Ashworth, J.R., Hutchison, R.: Water in non-carbonaceous stony meteorites. Nature **256**, 714 (1975)
12. Atreya, S.K., Baines, K.H., Egeler, P.A.: An ocean of water-ammonia on Neptune and Uranus: clues from tropospheric cloud structure. In: Bulletin of the American Astronomical Society, vol. 38, p. 489 (2006)
13. Aumann, H.H., Beichman, C.A., Gillett, F.C., de Jong, T., Houck, J.R., Low, F.J., Neugebauer, G., Walker, R.G., Wesselius, P.R.: Discovery of a shell around Alpha Lyrae. Astrophys. J. Lett. **278**, L23–L27 (1984)
14. Babkovskaia, N., Poutanen, J.: A self-consistent model of a 22 GHz water maser in a dusty environment near late-type stars. Astron. Astrophys. **447**, 949–962 (2006)
15. Bailey, J.: A comparison of water vapor line parameters for modeling the Venus deep atmosphere. Icarus **201**, 444–453 (2009)
16. Barkume, K.M., Brown, M.E., Schaller, E.L.: Water ice on the satellite of Kuiper belt object 2003 EL_{61}. Astrophys. J. Lett. **640**, L87–L89 (2006)
17. Barlow, M.J., Nguyen-Q-Rieu, Truong-Bach, Cernicharo, J., Gonzalez-Alfonso, E., Liu, X.-W., Cox, P., Sylvester, R.J., Clegg, P.E., Griffin, M.J., Swinyard, B.M., Unger, S.J.,

DOI 10.1007/978-90-481-9984-6,

Baluteau, J.-P., Caux, E., Cohen, M., Cohen, R.J., Emery, R.J., Fischer, J., Furniss, I., Glencross, W.M., Greenhouse, M.A., Gry, C., Joubert, M., Lim, T., Lorenzetti, D., Nisini, B., Omont, A., Orfei, R., Pequignot, D., Saraceno, P., Serra, G., Skinner, C.J., Smith, H.A., Walker, H.J., Armand, C., Burgdorf, M., Ewart, D., di Giorgio, A., Molinari, S., Price, M., Sidher, S., Texier, D., Trams, N.: The rich far-infrared water vapour spectrum of W Hya. Astron. Astrophys. **315**, L241–L244 (1996)

18. Barnes, J.W., Brown, R.H., Soderblom, L., Sotin, C., Lemouelic, S., Rodriguez, S., Lecorre, L., Buratti, B.J., Pitman, K.M., Clark, R.N., Jaumann, R., Hayne, P.: Titan as an icy moon: evidence for cryovolcanism and tectonics from Cassini/VIMS. LPI Contrib. **1357**, 13–15 (2007)
19. Barucci, M.A., Fulchignoni, M., Lazzarin, M.: Water ice in primitive asteroids? Planet. Space Sci. **44**, 1047–1049 (1996)
20. Barucci, M.A., Cruikshank, D.P., Dotto, E., Merlin, F., Poulet, F., Dalle Ore, C., Fornasier, S., de Bergh, C.: Nitrogen and methane ices on the surface of Sedna? In: Bulletin of the American Astronomical Society, vol. 37, p. 744 (2005)
21. Basri, G., Ramos-Stierle, F., Soto, K., Lewis, T., Reiners, A., Borucki, W., Koch, D.: The Kepler mission: terrestrial extrasolar planets and stellar activity. In: van Belle, G. (ed.) 14th Cambridge Workshop on Cool Stars, Stellar Systems, and the Sun. Astronomical Society of the Pacific Conference Series, vol. 384, p. 281 (2008)
22. Bauer, S.J.: Water on Venus—lack or loss? Ann. Geophys. **1**, 477–479 (1983)
23. Bergin, E.A., Lellouch, E., Harwit, M., Gurwell, M.A., Melnick, G.J., Ashby, M.L.N., Chin, G., Erickson, N.R., Goldsmith, P.F., Howe, J.E., Kleiner, S.C., Koch, D.G., Neufeld, D.A., Patten, B.M., Plume, R., Schieder, R., Snell, R.L., Stauffer, J.R., Tolls, V., Wang, Z., Winnewisser, G., Zhang, Y.F.: Submillimeter wave astronomy satellite observations of Jupiter and Saturn: detection of 557 GHz water emission from the upper atmosphere. Astrophys. J. Lett. **539**, L147–L150 (2000)
24. Bernath, P.F.: Water in sunspots and stars. Highlights Astron. **12**, 70–72 (2002)
25. Bertaux, J.-L., Vandaele, A.-C., Korablev, O., Villard, E., Fedorova, A., Fussen, D., Quémerais, E., Belyaev, D., Mahieux, A., Montmessin, F., Muller, C., Neefs, E., Nevejans, D., Wilquet, V., Dubois, J.P., Hauchecorne, A., Stepanov, A., Vinogradov, I., Rodin, A., Bertaux, J.-L., Nevejans, D., Korablev, O., Montmessin, F., Vandaele, A.-C., Fedorova, A., Cabane, M., Chassefière, E., Chaufray, J.Y., Dimarellis, E., Dubois, J.P., Hauchecorne, A., Leblanc, F., Lefèvre, F., Rannou, P., Quémerais, E., Villard, E., Fussen, D., Muller, C., Neefs, E., van Ransbeeck, E., Wilquet, V., Rodin, A., Stepanov, A., Vinogradov, I., Zasova, L., Forget, F., Lebonnois, S., Titov, D., Rafkin, S., Durry, G., Gérard, J.C., Sandel, B.: A warm layer in Venus' cryosphere and high-altitude measurements of HF, HCl, H_2O and HDO. Nature **450**, 646–649 (2007)
26. Bertone, G., Hooper, D., Silk, J.: Particle dark matter: evidence, candidates and constraints. Phys. Rep. **405**, 279–390 (2005)
27. Bertout, C., Basri, G., Bouvier, J.: Accretion disks around T Tauri stars. Astrophys. J. **330**, 350–373 (1988)
28. Beuther, H., Walsh, A., Schilke, P., Sridharan, T.K., Menten, K.M., Wyrowski, F.: CH_3OH and H_2O masers in high-mass star-forming regions. Astron. Astrophys. **390**, 289–298 (2002)
29. Bézard, B., Tsang, C.C.C., Carlson, R.W., Piccioni, G., Marcq, E., Drossart, P.: Water vapor abundance near the surface of Venus from Venus Express/VIRTIS observations. J. Geophys. Res. (Planets) **114**, E00B39 (2009)
30. Bjoraker, G.L., Larson, H.P., Kunde, V.G.: The abundance and distribution of water vapor in Jupiter's atmosphere. Astrophys. J. **311**, 1058–1072 (1986)
31. Bjoraker, G.L., Herter, T., Gull, G., Stolovy, S., Pirger, B.: Detection of water int he fireball of fragments G and K of Comet Shoemaker-Levy 9. In: Bulletin of the American Astronomical Society, vol. 26, p. 1578 (1994)
32. Bjoraker, G.L., Stolovy, S.R., Herter, T.L., Gull, G.E., Pirger, B.E.: Detection of water after the collision of fragments G and K of Comet Shoemaker-Levy 9 with Jupiter. Icarus **121**, 411–421 (1996)

33. Bjoraker, G., Achterberg, R., Anderson, C., Samuelson, R., Carlson, R., Jennings, D.: Cassini/CIRS observations of water vapor in Titan's stratosphere. In: Bulletin of the American Astronomical Society, vol. 40, p. 448 (2008)
34. Bobylev, V.V.: Searching for stars closely encountering with the Solar system. ArXiv e-prints (2010)
35. Bockelée-Morvan, D., Woodward, C.E., Kelley, M.S., Wooden, D.H.: Water in comets 71P/Clark and C/2004 B1 (linear) with Spitzer. Astrophys. J. **696**, 1075–1083 (2009)
36. Bohor, B.F.: In: Word, P.D., Sharpton, V.L. (eds.) Global Catastrophes in Earth History. Geol. Soc. Am., Boulder (1990)
37. Bonev, T., Borisov, G., Ivanova, A.: H_2O^+ ions in the plasma tail of comet Ikeya-Zhang. Publ. Obs. Astron. Beogr. **73**, 73–80 (2002)
38. Bono, P., Boni, C.: Water supply of Rome in antiquity and today. Environ. Geol. **27**, 126–134 (1996)
39. Böttcher, T., Huber, L., Le Corre, L., Leitner, J., McCarthy, D., Nilsson, R., Teixeira, C., Vaquer Araujo, S., Wilson, R.C., Adjali, F., Altenburg, M., Briani, G., Buchas, P., Le Postollec, A., Meier, T.: The HADES mission concept—astrobiological survey of Jupiter's icy moon Europa. Int. J. Astrobiol. **8**, 321–329 (2009)
40. Bounamam, C., von Bloh, W., Franck, S.: How rare is complex life in the Milky Way? Astrobiology **7**, 745–756 (2007)
41. Bourgeois, J., Wiberg, P.L., Hansen, T.A.: Sedimentological effects of tsunamis, with particular reference to impact-generated and volcanogenic waves. LPI Contrib. **673**, 21 (1988)
42. Boynton, W.V., Feldman, W.C., Mitrofanov, I., Evans, L.G., Reedy, R.C., Squyres, S.W., Starr, R., Trombka, J.I., D'Uston, C., Arnold, J.R., Englert, P.A.J., Metzger, A.E., Wänke, H., Brückner, J., Drake, D.M., Shinohara, C., Fellows, C., Hamara, D.K., Harshman, K.: Early results of the Mars Odyssey gamma-ray spectrometer (GRS): ice and other cool stuff. Meteorit. Planet. Sci. Suppl. **37**, 21 (2002)
43. Braatz, J.A., Gugliucci, N.E.: The discovery of water maser emission from eight nearby galaxies. Astrophys. J. **678**, 96–101 (2008)
44. Brand, J., Cesaroni, R., Caselli, P., Catarzi, M., Codella, C., Comoretto, G., Curioni, G.P., Curioni, P., di Franco, S., Felli, M., Giovanardi, C., Olmi, L., Palagi, F., Palla, F., Panella, D., Pareschi, G., Rossi, E., Speroni, N., Tofani, G.: The Arcetri catalogue of H_2O maser sources update. Astron. Astrophys. Suppl. **103**, 541–572 (1994)
45. Brecht, S.H., Ledvina, S.A.: The loss of water from Mars: numerical results and challenges. Icarus **206**, 164–173 (2010)
46. Brown, M.E., Calvin, W.M.: Evidence for crystalline water and ammonia ices on Pluto's satellite Charon. Science **287**, 107–109 (2000)
47. Brown, M.E., Barkume, K.M., Ragozzine, D., Schaller, E.L.: A collisional family of icy objects in the Kuiper belt. Nature **446**, 294–296 (2007)
48. Brown, R.H., Cruikshank, D.P., Pendleton, Y.: Water ice on Kuiper belt object 1996 TO_66. Astrophys. J. Lett. **519**, L101–L104 (1999)
49. Brown, T.M.: Transmission spectra as diagnostics of extrasolar giant planet atmospheres. Astrophys. J. **553**, 1006–1026 (2001)
50. Buitrago, J., Mediavilla, E.: Dynamics of the dust material surrounding VEGA. Astron. Astrophys. **148**, L8–L10 (1985)
51. Burke, B.F., Papa, D.C., Papadopoulos, G.D., Schwartz, P.R., Knowles, S.H., Sullivan, W.T., Meeks, M.L., Moran, J.M.: Studies of H_2O sources by means of a very-long interferometer. Astrophys. J. Lett. **160**, L63 (1970)
52. Busso, M., Gallino, R., Wasserburg, G.J.: Nucleosynthesis in asymptotic giant branch stars: relevance for galactic enrichment and solar system formation. Ann. Rev. Astron. Astrophys. **37**, 239–309 (1999)
53. Butler, B.J., Slade, M.A., Muhleman, D.O.: The nature of the mercury polar radar features. In: Robinbson, M., Taylor, G.J. (eds.) Workshop on Mercury: Space Environment, Surface, and Interior, p. 9 (2001)
54. Campbell, D.B., Campbell, B.A., Carter, L.M., Margot, J.-L., Stacy, N.J.S.: No evidence for thick deposits of ice at the lunar south pole. Nature **443**, 835–837 (2006)

55. Campins, H., Hargrove, K., Howell, E.S., Kelley, M.S., Licandro, J., Mothé-Diniz, T., Ziffer, J., Fernandez, Y., Pinilla-Alonso, N.: Confirming water ice on the surface of asteroid 24 Themis. In: AAS/Division for Planetary Sciences Meeting Abstracts, vol. 41, p. 32.05 (2009)
56. Carbon, D.F., Goorvitch, D.: Spectrum synthesis of hot water in sunspots and selected cool stars. In: Bulletin of the American Astronomical Society, vol. 28, p. 1382 (1996)
57. Carpenter, J.M., Bouwman, J., Mamajek, E.E., Meyer, M.R., Hillenbrand, L.A., Backman, D.E., Henning, T., Hines, D.C., Hollenbach, D., Kim, J.S., Moro-Martin, A., Pascucci, I., Silverstone, M.D., Stauffer, J.R., Wolf, S.: Formation and evolution of planetary systems: properties of debris dust around solar-type stars. Astrophys. J. Suppl. **181**, 197–226 (2009)
58. Carr, M.H.: Water on Mars. Nature **326**, 30–35 (1987)
59. Carr, M.H., Head, J.W.: Geologic history of Mars. Earth Planet. Sci. Lett. **294**, 185–203 (2010)
60. Cassen, P., Reynolds, R.T., Peale, S.J.: Is there liquid water on Europa. Geophys. Res. Lett. **6**, 731–734 (1979)
61. Castillo-Rogez, J.: Internal structure of Rhea. J. Geophys. Res. (Planets) **111**, 11005 (2006)
62. Castillo-Rogez, J.C., Matson, D.L., Sotin, C., Johnson, T.V., Lunine, J.I., Thomas, P.C.: Iapetus' geophysics: rotation rate, shape, and equatorial ridge. Icarus **190**, 179–202 (2007)
63. Catelan, M.: Structure and evolution of low-mass stars: an overview and some open problems. In: Roig, F., Lopes, D. (eds.) Graduate School in Astronomy: XI Special Courses at the National Observatory of Rio de Janeiro (XI CCE). American Institute of Physics Conference Series, vol. 930, pp. 39–90 (2007)
64. Cavicchioli, R.: Extremophiles and the search for extraterrestrial life. Astrobiology **2**, 281–292 (2002)
65. Chaussidon, M.: Planetary science: the early Moon was rich in water. Nature **454**, 170–172 (2008)
66. Chen, E.M.A., Nimmo, F.: Thermal and orbital evolution of Tethys as constrained by surface observations. In: Lunar and Planetary Institute Conference Abstracts, vol. 39, p. 1968 (2008)
67. Cheung, A.C., Rank, D.M., Townes, C.H., Thornton, D.D., Welch, W.J.: Delection of water in interstellar regions by its microwave radiation. Nature **221**, 626 (1969)
68. Chyba, C.F., Hand, K.P.: ASTROBIOLOGY: the study of the living Universe. Ann. Rev. Astron. Astrophys. **43**, 31–74 (2005)
69. Clark, R.N., Curchin, J.M., Jaumann, R., Cruikshank, D.P., Brown, R.H., Hoefen, T.M., Stephan, K., Moore, J.M., Buratti, B.J., Baines, K.H., Nicholson, P.D., Nelson, R.M.: Compositional mapping of Saturn's satellite Dione with Cassini VIMS and implications of dark material in the Saturn system. Icarus **193**, 372–386 (2008)
70. Claussen, M.J., Goss, W.M., Frail, D.A., Seta, M.: No water masers associated with supernova remnants. Astron. J. **117**, 1387–1391 (1999)
71. Codella, C., Palla, F.: The nature of massive protostellar candidates. I. A search for water masers towards color-selected IRAS sources. Astron. Astrophys. **302**, 528 (1995)
72. Codella, C., Felli, M., Natale, V.: The occurrence of H_2O masers in the early stages of star formation. Astron. Astrophys. **311**, 971–980 (1996)
73. Cohen, N.L., Malkan, M.A.: A search for H_2O maser emission from globular clusters. Astron. J. **84**, 74–76 (1979)
74. Cohen, N.L., Malkan, M.A., Dickey, J.M.: A passive SETI in globular clusters at the hydroxyl and water lines. Icarus **41**, 198–204 (1980)
75. Connerney, J.E.P., Waite, J.H.: New model of Saturn's ionosphere with an influx of water from the rings. Nature **312**, 136–138 (1984)
76. Conway, S.J., Balme, M.R., Towner, M.C., Murray, J.B.: Can water move sediment on present-day Mars? Insights into gully formation from laboratory simulations. In: Lunar and Planetary Institute Science Conference Abstracts, vol. 41, p. 1894 (2010)
77. Cordell, B.M.: Martian climatic change—a magnetic trigger. Geophys. Res. Lett. **7**, 1065–1068 (1980)
78. Cosmovici, C.B., Montebugnoli, S., Orfei, A., Pogrebenko, S., Cortiglioni, S.: The puzzling detection of the 22 GHz water emission line in Comet Hyakutake at perihelion. Planet. Space Sci. **46**, 467–470 (1998)

79. Cosmovici, C.B., Pogrebenko, S., Montebugnoli, S., Maccaferri, G.: Radio searching for extrasolar cometary impacts at 22 Ghz (water MASER emission). In: Catastrophic Events and Mass Extinctions: Impacts and Beyond, p. 3002 (2001)
80. Cosmovici, C.B., Montebugnoli, S., Righini, S., Salerno, E., Zoni, L., Flamini, E.: The search of water in exoplanets. In: European Planetary Science Congress 2006, p. 80 (2006)
81. Coustenis, A., Schneider, J., Bockelée-Morvan, D., Rauer, H., Wittemberg, R., Chassefière, E., Greene, T., Penny, A., Guillot, T.: Spectroscopy of 51 Peg B: search for atmospheric signatures. In: Soderblom, D. (ed.) Planets Beyond the Solar System and the Next Generation of Space Missions. Astronomical Society of the Pacific Conference Series, vol. 119, p. 101 (1997)
82. Coustenis, A., Salama, A., Lellouch, E., Encrenaz, T., de Graauw, T., Bjoraker, G.L., Samuelson, R.E., Gautier, D., Feuchtgruber, H., Kessler, M.F., Orton, G.S.: Titan's atmosphere from Iso observations: temperature, composition and detection of water vapor. In: Bulletin of the American Astronomical Society, vol. 30, p. 1060 (1998)
83. Crawford, G.D., Stevenson, D.J.: Gas-driven water volcanism and resurfacing on Europa. In: Lunar and Planetary Institute Conference Abstracts, vol. 16, pp. 148–149 (1985)
84. Crawford, G.D., Stevenson, D.J.: Gas-driven water volcanism in the resurfacing of Europa. Icarus **73**, 66–79 (1988)
85. Crifo, J.F., Slanina, Z.: The formation and role of water dimers in a cometary atmosphere, with application to Comet P/Halley. Astrophys. J. **383**, 351–355 (1991)
86. Crouch, D.P.: Environmental geology of ancient Greek cities. Environ. Geol. **27**, 233–245 (1996)
87. Cruikshank, D.P., Schmitt, B., Roush, T.L., Owen, T.C., Quirico, E., Geballe, T.R., de Bergh, C., Bartholomew, M.J., Dalle Ore, C.M., Douté, S., Meier, R.: Water ice on Triton. Icarus **147**, 309–316 (2000)
88. Currie, T., Plavchan, P., Kenyon, S.J.: A Spitzer study of debris disks in the young nearby cluster NGC 2232: icy planets are common around $\sim$1.5–3 M_{solar} stars. Astrophys. J. **688**, 597–615 (2008)
89. Czechowski, L., Leliwa-Kopystyński, J.: The Iapetus's ridge: possible explanations of its origin. Adv. Space Res. **42**, 61–69 (2008)
90. Dahmani, R., Khanna, R.K.: H_2O on Io? IR spectra of SO_2/H_2O mixed ices in the 5000–450 cm^{-1} region. Astrophys. Space Sci. **236**, 125–133 (1996)
91. Dalle Ore, C.M., Cruikshank, D.P., Owen, T.C., Geballe, T.R., Roush, T.L., Khare, B.N., de Bergh, C.: Surface composition of RHEA from reflectance spectra, 0.2 to 3.6 MU M. In: AAS/Division for Planetary Sciences Meeting Abstracts, vol. 31, p. 03.01 (1999)
92. Darling, J., Brogan, C., Johnson, K.: Ubiquitous water masers in nearby star-forming galaxies. Astrophys. J. Lett. **685**, L39–L42 (2008)
93. Darling, J.K., Brogan, C., Johnson, K.: The discovery of water masers in nearby star-forming galaxies. In: American Astronomical Society Meeting Abstracts, vol. 213, p. 445.01 (2009)
94. Dauphas, N., Robert, F., Marty, B.: The late asteroidal and cometary bombardment of Earth as recorded in water deuterium to protium ratio. Icarus **148**, 508–512 (2000)
95. Davies, J.H.: Did a mega-collision dry Venus' interior? Earth Planet. Sci. Lett. **268**, 376–383 (2008)
96. Davies, J.K., Roush, T.L., Cruikshank, D.P., Bartholomew, M.J., Geballe, T.R., Owen, T., de Bergh, C.: The detection of water ice in Comet Hale-Bopp. Icarus **127**, 238–245 (1997)
97. Davoust, E.: Silence at the water hole. NASA STI/Recon Technical Report A, vol. 89, p. 27056 (1988)
98. Davoust, E.: The cosmic water hole. NASA STI/Recon Technical Report A, vol. 91, p. 31098 (1991)
99. de Bergh, C.: Kuiper belt: water and organics. In: Ehrenfreund, P. et al. (eds.) Astrobiology: Future Perspectives. Astrophysics and Space Science Library, vol. 305, p. 205. Kluwer Academic, Dordrecht (2004)
100. de Jong, T.: Water masers in a protostellar gas cloud. Astron. Astrophys. **26**, 297 (1973)
101. Debes, J.H., Weinberger, A.J., Schneider, G.: Complex organic materials in the circumstellar disk of HR 4796A. Astrophys. J. Lett. **673**, L191–L194 (2008)

102. Dehant, V., Lammer, H., Kulikov, Y.N., Grießmeier, J.-M., Breuer, D., Verhoeven, O., Karatekin, Ö., van Hoolst, T., Korablev, O., Lognonné, P.: Planetary magnetic dynamo effect on atmospheric protection of early Earth and Mars. Space Sci. Rev. **129**, 279–300 (2007)
103. Deloule, E., Robert, F.: Origin of water in meteorites: ion-probe determinations of D/h ratios in chondrules. In: Lunar and Planetary Institute Conference Abstracts, vol. 27, p. 307 (1996)
104. Deloule, E., Robert, F., Doukhan, J.C.: Interstellar hydroxyl in meteoritic chondrules: implications for the origin of water in the inner solar system. Geochim. Cosmoch. Acta **62**, 3367–3378 (1998)
105. Delva, M., Volwerk, M., Mazelle, C., Chaufray, J.Y., Bertaux, J.L., Zhang, T.L., Vörös, Z.: Hydrogen in the extended Venus exosphere. Geophys. Res. Lett. **36**, 1203 (2009)
106. Dohm, J.M., Anderson, R.C., Barlow, N.G., Miyamoto, H., Davies, A.G., Jeffrey Taylor, G., Baker, V.R., Boynton, W.V., Keller, J., Kerry, K., Janes, D., Fairén, A.G., Schulze-Makuch, D., Glamoclija, M., Marinangeli, L., Ori, G.G., Strom, R.G., Williams, J.-P., Ferris, J.C., Rodríguez, J.A.P., de Pablo, M.A., Karunatillake, S.: Recent geological and hydrological activity on Mars: The Tharsis/Elysium corridor. Planet. Space Sci. **56**, 985–1013 (2008)
107. Donahue, T.M., Hoffman, J.H., Hodges, R.R., Watson, A.J.: Venus was wet—a measurement of the ratio of deuterium to hydrogen. Science **216**, 630–633 (1982)
108. Donahue, T.M., Grinspoon, D.H., Hartle, R.E., Hodges, R.R. Jr.: Ion/neutral escape of hydrogen and deuterium: evolution of water. In: Bougher, S.W., Hunten, D.M., Philips, R.J. (eds.) Venus II: Geology, Geophysics, Atmosphere, and Solar Wind Environment, p. 385 (1997)
109. Dorofeeva, V.A., Ruskol, E.L.: On the thermal history of Saturn's satellites Titan and Enceladus. Sol. Syst. Res. **44**, 192–201 (2010)
110. Dulieu, F., Amiaud, L., Congiu, E., Fillion, J.-H., Matar, E., Momeni, A., Pirronello, V., Lemaire, J.L.: Experimental evidence for water formation on interstellar dust grains by hydrogen and oxygen atoms. Astron. Astrophys. **512**, 30 (2010)
111. Doyle, D., Pilbratt, G., Tauber, J.: The Herschel and Planck space telescopes. IEEE Proc. **97**, 1403–1411 (2009)
112. Dybczyński, P.A., Leto, G., Jakubík, M., Paulech, T., Neslušan, L.: The simulation of the outer Oort cloud formation. The first giga-year of the evolution. Astron. Astrophys. **487**, 345–355 (2008)
113. Edmonds, R., Wagg, J., Momjian, E., Carilli, C.L., Wilner, D.J., Humphreys, E.M.L., Menten, K.M., Hughes, D.H.: An EVLA search for water megamaser emission in the submm galaxy SMM J16359+6612 at $z = 2.5$. ArXiv e-prints (2008)
114. Ehrenfreund, P., Irvine, W., Becker, L., Blank, J., Brucato, J.R., Colangeli, L., Derenne, S., Despois, D., Dutrey, A., Fraaije, H., Lazcano, A., Owen, T., Robert, F. (International Space Science Institute ISSI-Team): Astrophysical and astrochemical insights into the origin of life. Rep. Prog. Phys. **65**, 1427–1487 (2002)
115. Ehrenreich, D., Hébrard, G., Lecavelier des Etangs, A., Sing, D.K., Désert, J.-M., Bouchy, F., Ferlet, R., Vidal-Madjar, A.: A Spitzer search for water in the transiting exoplanet HD 189733b. Astrophys. J. Lett. **668**, L179–L182 (2007)
116. Elder, C., Helfenstein, P., Thomas, P., Veverka, J., Burns, J.A., Denk, T., Porco, C.: Tethys' mysterious equatorial band. In: Bulletin of the American Astronomical Society, vol. 38, p. 429 (2007)
117. Elitzur, M.: A dynamical explanation for the high water abundance detected in Orion. Astrophys. J. **229**, 560–566 (1979)
118. Elitzur, M.: Masers in the interstellar medium. In: Hollenbach, D.J., Thronson, H.A. Jr. (eds.) Interstellar Processes. Astrophysics and Space Science Library, vol. 134, pp. 763–780 (1987)
119. Elitzur, M., Hollenbach, D.J., McKee, C.F.: H_2O masers in star-forming regions. Astrophys. J. **346**, 983–990 (1989)
120. Elkins-Tanton, L.T., Seager, S.: Ranges of atmospheric mass and composition of Super-Earth exoplanets. Astrophys. J. **685**, 1237–1246 (2008)
121. Elmegreen, B.G., Morris, M.: Disk structure in protostellar H_2O maser sources. Astrophys. J. **229**, 593–603 (1979)
122. England, C.: Are there oceans under the ice of small Saturnian and Uranian moons? In: Bulletin of the American Astronomical Society, vol. 35, p. 939 (2003)

123. Fairén, A.G.: A cold and wet Mars. Icarus **208**, 165–175 (2010)
124. Fairén, A.G., Schulze-Makuch, D., Rodríguez, A.P., Fink, W., Davila, A.F., Uceda, E.R., Furfaro, R., Amils, R., McKay, C.P.: Evidence for Amazonian acidic liquid water on Mars–a reinterpretation of MER mission results. Planet. Space Sci. **57**, 276–287 (2009)
125. Fan, C., Schulze-Makuch, D., Xie, H., Lu, N.: Investigation of water signatures at gully-exposed sites on Mars by hyperspectral image analysis. Planet. Space Sci. **57**, 93–104 (2009)
126. Fanale, F.P., Salvail, J.R.: The water regime of asteroid (1) Ceres. Icarus **82**, 97–110 (1989)
127. Fassett, C.I., Dickson, J.L., Head, J.W., Levy, J.S., Marchant, D.R.: Supraglacial and proglacial valleys on Amazonian Mars. Icarus **208**, 86–100 (2010)
128. Fedorova, A.A., Trokhimovsky, S., Korablev, O., Montmessin, F.: Viking observation of water vapor on Mars: revision from up-to-date spectroscopy and atmospheric models. Icarus **208**, 156–164 (2010)
129. Fedorova, A., Korablev, O., Vandaele, A.-C., Bertaux, J.-L., Belyaev, D., Mahieux, A., Neefs, E., Wilquet, W.V., Drummond, R., Montmessin, F., Villard, E.: HDO and H_2O vertical distributions and isotopic ratio in the Venus mesosphere by Solar occultation at infrared spectrometer on board Venus express. J. Geophys. Res. (Planets) **113**, E00B22 (2008)
130. Felli, M., Brand, J., Cesaroni, R., Codella, C., Comoretto, G., di Franco, S., Massi, F., Moscadelli, L., Nesti, R., Olmi, L., Palagi, F., Panella, D., Valdettaro, R.: Water maser variability over 20 years in a large sample of star-forming regions: the complete database. Astron. Astrophys. **476**, 373–664 (2007)
131. Fenton, L.K., Geissler, P.E., Haberle, R.M.: Global warming on Mars. In: AGU Fall Meeting Abstracts, p. A47 (2006)
132. Ferrière, K.M.: The interstellar environment of our galaxy. Rev. Mod. Phys. **73**, 1031–1066 (2001)
133. Ford, E.B., Colon, K.D., Blake, C., Lee, B., Mahadevan, S.: First results of exoplanet observations with the Gran telescopio canarias: narrow-band transit photometry capable of detecting super-earth-size planets. In: Bulletin of the American Astronomical Society, vol. 41, p. 287 (2010)
134. Forterre, P.: Defining life: the virus viewpoint. Orig. Life Evol. Biosph. **40**, 151–160 (2010)
135. Fulvio, D., Guglielmino, S., Favone, T., Palumbo, M.E.: Near-infrared laboratory spectra of H_2O trapped in N_2, CH_4, and CO: hints for trans-Neptunian objects' observations. Astron. Astrophys. **511**, 62 (2010)
136. Furuya, R.S., Kitamura, Y., Wootten, A., Claussen, M.J., Kawabe, R.: Water maser survey toward low-mass young stellar objects in the northern sky with the Nobeyama 45 meter telescope and the very large array. Astrophys. J. Suppl. **144**, 71–134 (2003)
137. Gaidos, E.J., Nimmo, F.: Planetary science: tectonics and water on Europa. Nature **405**, 637 (2000)
138. Gayon, J.: Defining life: synthesis and conclusions. Orig. Life Evol. Biosph. **40**, 231–244 (2010)
139. Gibb, E.L., Whittet, D.C.B., Schutte, W.A., Boogert, A.C.A., Chiar, J.E., Ehrenfreund, P., Gerakines, P.A., Keane, J.V., Tielens, A.G.G.M., van Dishoeck, E.F., Kerkhof, O.: An inventory of interstellar ices toward the embedded protostar W33A. Astrophys. J. **536**, 347–356 (2000)
140. Gibbard, S., Levy, E.H., Lunine, J.I.: Generation of lightning in Jupiter's water cloud. Nature **378**, 592–595 (1995)
141. Gierasch, P.J., Ingersoll, A.P., Banfield, D., Ewald, S.P., Helfenstein, P., Simon-Miller, A., Vasavada, A., Breneman, H.H., Senske, D.A. (Galileo Imaging Team): Observation of moist convection in Jupiter's atmosphere. Nature **403**, 628–630 (2000)
142. Glassgold, A.E., Meijerink, R., Najita, J.R.: Formation of water in the warm atmospheres of protoplanetary disks. Astrophys. J. **701**, 142–153 (2009)
143. Goguen, J.D., Orzechowska, G.E., Johnson, P.V., Tsapin, A.I., Kanik, I.: UV photolysis of amino acids in water ice: how long can they survive on Europa? In: Mackwell, S., Stansbery, E. (eds.) 37th Annual Lunar and Planetary Science Conference, vol. 37, p. 2006 (2006)
144. Gomes, R.D.S.: On the origin of the Kuiper belt. Celest. Mech. Dyn. Astron. **104**, 39–51 (2009)

145. Greenberg, R., Hoppa, G.V., Tufts, B.R., Geissler, P., Riley, J.: Chaos, cracks and ridges: surface effects of thin ice over liquid water on Europa. In: Lunar and Planetary Institute Conference Abstracts, vol. 30, p. 1421 (1999)
146. Grundy, W.M., Young, L.A., Young, E.F.: Discovery of CO_2 ice and leading-trailing spectral asymmetry on the Uranian satellite Ariel. Icarus **162**, 222–229 (2003)
147. Grundy, W.M., Young, L.A., Spencer, J.R., Johnson, R.E., Young, E.F., Buie, M.W.: Distributions of H_2O and CO_2 ices on Ariel, Umbriel, Titania, and Oberon from IRTF/SpeX observations. Icarus **184**, 543–555 (2006)
148. Güdel, M.: The Sun in time: activity and environment. Living Rev. Sol. Phys. **4**, 3 (2007)
149. Guedel, M.: X-Rays, dust, ice, and gas in time (X-DIGIT). In: XMM-Newton Proposal ID #06523509, p. 89 (2009)
150. Guillot, T.: Condensation of methane ammonia and water in the inhibition of convection in giant planets. Science **269**, 1697 (1995)
151. Haberle, R.M.: History and progress of GCM simulations on recent Mars climate change. In: Mackwell, S., Stansbery, E. (eds.) Lunar and Planetary Institute Conference Abstracts, vol. 35, p. 2010 (2004)
152. Hahn, J.M., Malhotra, R.: Neptune's migration into a stirred-up Kuiper belt: a detailed comparison of simulations to observations. Astron. J. **130**, 2392–2414 (2005)
153. Hansen, C.J., Esposito, L., Stewart, A.I.F., Colwell, J., Hendrix, A., Pryor, W., Shemansky, D., West, R.: Enceladus' water vapor plume. Science **311**, 1422–1425 (2006)
154. Hansen, C.J., Esposito, L.W., Stewart, A.I.F., Meinke, B., Wallis, B., Colwell, J.E., Hendrix, A.R., Larsen, K., Pryor, W., Tian, F.: Water vapour jets inside the plume of gas leaving Enceladus. Nature **456**, 477–479 (2008)
155. Hanslmeier, A.: Solar variations and climate on planets. In: Demircan, O., Selam, S.O., Albayrak, B. (eds.) Solar and Stellar Physics Through Eclipses. Astronomical Society of the Pacific Conference Series, vol. 370, p. 30 (2007)
156. Hanslmeier, A.: Habitability and Cosmic Catastrophes. Springer, Berlin (2009)
157. Hartmann, L., Calvet, N., Gullbring, E., D'Alessio, P.: Accretion and the evolution of T Tauri disks. Astrophys. J. **495**, 385 (1998)
158. Hartogh, P., Lellouch, E., Crovisier, J., Banaszkiewicz, M., Bensch, F., Bergin, E.A., Billebaud, F., Biver, N., Blake, G.A., Blecka, M.I., Blommaert, J., Bockelée-Morvan, D., Cavalié, T., Cernicharo, J., Courtin, R., Davis, G., Decin, L., Encrenaz, P., Encrenaz, T., González, A., de Graauw, T., Hutsemékers, D., Jarchow, C., Jehin, E., Kidger, M., Küppers, M., de Lange, A., Lara, L.-M., Lis, D.C., Lorente, R., Manfroid, J., Medvedev, A.S., Moreno, R., Naylor, D.A., Orton, G., Portyankina, G., Rengel, M., Sagawa, H., Sánchez-Portal, M., Schieder, R., Sidher, S., Stam, D., Swinyard, B., Szutowicz, S., Thomas, N., Thornhill, G., Vandenbussche, B., Verdugo, E., Waelkens, C., Walker, H.: Water and related chemistry in the solar system. A guaranteed time key programme for Herschel. Planet. Space Sci. **57**, 1596–1606 (2009)
159. Haruyama, J., Ohtake, M., Matsunaga, T., Morota, T., Honda, C., Yokota, Y., Pieters, C.M., Hara, S., Hioki, K., Saiki, K., Miyamoto, H., Iwasaki, A., Abe, M., Ogawa, Y., Takeda, H., Shirao, M., Yamaji, A., Josset, J.-L.: Lack of exposed ice inside lunar south pole Shackleton Crater. Science **322**, 938 (2008)
160. Head, J.W., Solomon, S.C., McNutt, R.L., Blewett, D.T., Chapman, C.R., Domingue, D.L., Gillis-Davis, J.J., Hawkins, S.E., Helbert, J., Holsclaw, G.M., Izenberg, N.R., McClintock, W.E., Merline, W.J., Murchie, S.L., Phillips, R.J., Prockter, L.M., Robinson, M.S., Denevi, B.W., Sprague, A.L., Strom, R.G., Vilas, F., Watters, T.R., Zuber, M.T.: The MESSENGER mission to Mercury: new insights into geological processes and evolution from the first two encounters. In: Lunar and Planetary Institute Science Conference Abstracts. Lunar and Planetary Inst., Technical Report, vol. 40, p. 2198 (2009)
161. Hendrix, A.R., Hansen, C.J.: Iapetus: new results from Cassini Uvis. In: Lunar and Planetary Institute Science Conference Abstracts. Lunar and Planetary Inst., Technical Report, vol. 39, p. 2200 (2008)
162. Hibbitts, C.A., Hansen, G.B., McCord, T.B.: CO_2/water-ice/non-ice relationships on the Galilean satellites Ganymede and Callisto. In: Bulletin of the American Astronomical Society, vol. 34, p. 881 (2002)

163. Hirtzig, M., Tokano, T., Rodriguez, S., Le Mouélic, S., Sotin, C.: A review of Titan's atmospheric phenomena. Astron. Astrophys. Rev. **17**, 105–147 (2009)
164. Hodyss, R., Goguen, J.D., Johnson, P.V., Campbell, C., Kanik, I.: Release of N_2, CH_4, CO_2, and H_2O from surface ices on Enceladus. Icarus **197**, 152–156 (2008)
165. Hubbard, W.B., Burrows, A., Lunine, J.I.: Theory of giant planets. Ann. Rev. Astron. Astrophys. **40**, 103–136 (2002)
166. Huber, C., Waechtershaeuser, G.: Peptides by activation of amino acids with CO on (Ni, Fe)S surfaces: implications for the origin of life. Science **281**, 670 (1998)
167. Hueso, R., Sánchez-Lavega, A.: A three-dimensional model of moist convection for the giant planets II: Saturn's water and ammonia moist convective storms. Icarus **172**, 255–271 (2004)
168. Hui, L., Seager, S.: Atmospheric lensing and oblateness effects during an extrasolar planetary transit. Astrophys. J. **572**, 540–555 (2002)
169. Ingersoll, A.P., Pankine, A.A.: Subsurface heat transfer on Enceladus: conditions under which melting occurs. Icarus **206**, 594–607 (2010)
170. Irwin, J., Charbonneau, D., Nutzman, P., Falco, E.: The MEarth project: searching for transiting habitable super-Earth planets around nearby M-dwarfs. In: Stempels, E. (ed.) American Institute of Physics Conference Series, vol. 1094, pp. 445–448 (2009)
171. Jewitt, D.C., Luu, J.: Crystalline water ice on the Kuiper belt object (50000) Quaoar. Nature **432**, 731–733 (2004)
172. Johnson, N.M., Fegley, B.: Water on Venus: new insights from tremolite decomposition. Icarus **146**, 301–306 (2000)
173. Jones, B.W., Sleep, P.N., Underwood, D.R.: Habitability of known exoplanetary systems based on measured stellar properties. Astrophys. J. **649**, 1010–1019 (2006)
174. Jones, H.R.A., Pavlenko, Y., Viti, S., Tennyson, J.: Spectral analysis of water vapour in cool stars. Mon. Not. **330**, 675–684 (2002)
175. Jones, T.D., Lebofsky, L.A., Lewis, J.S., Marley, M.S.: The composition and origin of the C, P, and D asteroids—water as a tracer of thermal evolution in the outer belt. Icarus **88**, 172–192 (1990)
176. Jura, M.: Are there debris disks and Kuiper belt systems around first ascent red giants? In: Bulletin of the American Astronomical Society, vol. 37, p. 469 (2005)
177. Kaltenegger, L., Kasting, J.: Session 22. Habitability of super-Earths. Astrobiology **8**, 394–396 (2008)
178. Kargel, J.S., Strom, R.G.: Global climatic change on Mars. Sci. Am. **275**, 60–68 (1996)
179. Karlsson, H.R., Gibson, E.K., Clayton, R.N., Mayeda, T.K., Socki, R.A.: Extraterrestrial water of possible Martian origin in SNC meteorites: constraints from oxygen isotopes. In: Annual Meeting of the Meteoritical Society, p. 113 (1991)
180. Karlsson, H.R., Clayton, R.N., Gibson, E.K. Jr., Mayeda, T.K.: Water in SNC meteorites—evidence for a Martian hydrosphere. Science **255**, 1409–1411 (1992)
181. Kasting, J.F.: Runaway and moist greenhouse atmospheres and the evolution of earth and Venus. Icarus **74**, 472–494 (1988)
182. Kenyon, S.J., Bromley, B.C., O'Brien, D.P., Davis, D.R.: Formation and collisional evolution of Kuiper belt objects. In: The Solar System Beyond Neptune, pp. 293–313 (2008)
183. Kern, S.D., Elliot, J.L.: Discovery and characteristics of the Kuiper belt binary 2003QY90. Icarus **183**, 179–185 (2006)
184. Khurana, K.K., Kivelson, M.G., Russell, C.T.: Searching for liquid water in Europa by using surface observatories. Astrobiology **2**, 93–103 (2002)
185. Knacke, R.F., Noll, K.S., Brooke, T.Y., Geballe, T.R., Tokunaga, A.T.: Ground-based detection of water in Comet Halley. In: ESLAB Symposium on the Exploration of Halley's Comet. ESA Special Publication, vol. 250, pp. 99–101 (1986)
186. Knapp, G.R., Morris, M.: H_2O maser emission associated with T Tauri and other regions of star formation. Astrophys. J. **206**, 713–717 (1976)
187. Kobayashi, H., Ida, S., Tanaka, H.: The evidence of an early stellar encounter in Edgeworth Kuiper belt. Icarus **177**, 246–255 (2005)
188. Köhler, M., Mann, I., Li, A.: Complex organic materials in the HR 4796A disk? Astrophys. J. Lett. **686**, 95–98 (2008)

189. Kolovos, G., Varvoglis, H., Pylarinou, L.: Cometary water on Venus—impact modelling and possible evidence. Earth Moon Planets **54**, 103–117 (1991)
190. Kulikov, Y.N., Lammer, H., Lichtenegger, H.I.M., Terada, N., Ribas, I., Kolb, C., Langmayr, D., Lundin, R., Guinan, E.F., Barabash, S., Biernat, H.K.: Atmospheric and water loss from early Venus. Planet. Space Sci. **54**, 1425–1444 (2006)
191. Kuskov, O.L., Kronrod, V.A.: Internal structure of Callisto: evidence for subsurface ocean. In: 36th COSPAR Scientific Assembly. COSPAR, Plenary Meeting, vol. 36, p. 761 (2006)
192. Lammer, H., Kasting, J.F., Chassefière, E., Johnson, R.E., Kulikov, Y.N., Tian, F.: Atmospheric escape and evolution of terrestrial planets and satellites. Space Sci. Rev. **139**, 399–436 (2008)
193. Lammer, H., Bredehöft, J.H., Coustenis, A., Khodachenko, M.L., Kaltenegger, L., Grasset, O., Prieur, D., Raulin, F., Ehrenfreund, P., Yamauchi, M., Wahlund, J., Grießmeier, J.-M., Stangl, G., Cockell, C.S., Kulikov, Y.N., Grenfell, J.L., Rauer, H.: What makes a planet habitable? Astron. Astrophys. Rev. **17**, 181–249 (2009)
194. Larsen, K.W., Stewart, A., Barth, E., Toon, B., Shemansky, D.: Seasonal effects on the distribution of hydrocarbons and water in Titan's atmosphere. In: AGU Spring Meeting Abstracts, p. B3 (2006)
195. Lebofsky, L.A.: Asteroid 1 Ceres—evidence for water of hydration. Mon. Not. **182**, 17–21 (1978)
196. Lebofsky, L.A., Feierberg, M.A., Tokunaga, A.T., Larson, H.P., Johnson, J.R.: The 1.7- to 4.2-micron spectrum of asteroid 1 Ceres—evidence for structural water in clay minerals. Icarus **48**, 453–459 (1981)
197. Lebreton, J.-P., Coustenis, A., Lunine, J., Raulin, F., Owen, T., Strobel, D.: Results from the Huygens probe on Titan. Astron. Astrophys. Rev. **17**, 149–179 (2009)
198. Leisner, J.S., Russell, C.T., Dougherty, M.K., Blanco-Cano, X., Smith, E.J., Tsurutani, B.T.: Loss of water from Saturn's E-ring through ion pick-up. In: Mackwell, S., Stansbery, E. (eds.) 36th Annual Lunar and Planetary Science Conference. Lunar and Planetary Inst., Technical Report, vol. 36, p. 1935 (2005)
199. Leovy, C.B.: Martian dust storms and climate change. In: Carr, M., James, P., Conway, L., Pepin, R., Pollack, J. (eds.) MECA Workshop on the Evolution of the Martian Atmosphere. A Lunar and Planetary Institute Workshop held in Honolulu, Hawaii, August 9–10, 1985. LPI Technical Report 86-07, p. 26 (1986). Lunar and Planetary Institute, 3303 NASA Road 1, Houston, TX 77058
200. Levison, H.F., Morbidelli, A.: The formation of the Kuiper belt by the outward transport of bodies during Neptune's migration. Nature **426**, 419–421 (2003)
201. Lewis, S.R., Read, P.L.: Evidence for climate change on Mars. In: Solar System Update, p. 135 (2006)
202. Liang, M.-C., Yung, Y.L.: Modeling the distribution of H_2O and HDO in the upper atmosphere of Venus. J. Geophys. Res. (Planets) **114**, E00B28 (2009)
203. Lineweaver, C.H., Fenner, Y., Gibson, B.K.: The galactic habitable zone and the age distribution of complex life in the Milky Way. Science **303**, 59–62 (2004)
204. Litvak, M.M.: Hydroxyl and water masers in protostars. Science **165**, 855–861 (1969)
205. Lo, K.Y.: Mega-masers and galaxies. Ann. Rev. Astron. Astrophys. **43**, 625–676 (2005)
206. Lodders, K.: Jupiter formed with more tar than ice. Astrophys. J. **611**, 587–597 (2004)
207. Lodders, K., Fegley, B.: The Planetary Scientist's Companion. Oxford University Press, London (1998)
208. Lorenz, R., Stiles, B., Kirk, R.L., Allison, M.D., Persi del Marmo, P., Lunine, J.I.: Interpretation of Cassini radar measurement of Titan's spin: evidence of atmospheric angular momentum change and an internal water ocean. In: Bulletin of the American Astronomical Society, vol. 38, p. 544 (2007)
209. Luu, J.X., Jewitt, D.C., Trujillo, C.: Water ice in 2060 Chiron and its implications for Centaurs and Kuiper belt objects. Astrophys. J. Lett. **531**, 151–154 (2000)
210. Maercker, M., Schöier, F.L., Olofsson, H., Bergman, P., Ramstedt, S.: Circumstellar water vapour in M-type AGB stars: radiative transfer models, abundances, and predictions for HIFI. Astron. Astrophys. **479**, 779–791 (2008)

211. Maercker, M., Schöier, F.L., Olofsson, H., Bergman, P., Frisk, U., Hjalmarson, Å., Justtanont, K., Kwok, S., Larsson, B., Olberg, M., Sandqvist, A.: Circumstellar water vapour in M-type AGB stars: constraints from H{2}O(1{10}-1{01}) lines obtained with Odin. Astron. Astrophys. **494**, 243–252 (2009)
212. Maio, U., Ciardi, B., Dolag, K., Tornatore, L., Khochfar, S.: The transition from population III to population II-I star formation. ArXiv e-prints (2010)
213. Mallia, E.A., Blackwell, D.E., Petford, A.D.: Water vapour in sunspots. Nature **226**, 735–737 (1970)
214. Maltby, P., Avrett, E.H., Carlsson, M., Kjeldseth-Moe, O., Kurucz, R.L., Loeser, R.: A new sunspot umbral model and its variation with the solar cycle. Astrophys. J. **306**, 284–303 (1986)
215. Manoj, P., Bhatt, H.C.: Kinematics of Vega-like stars: lifetimes and temporal evolution of circumstellar dust disks. Astron. Astrophys. **429**, 525–530 (2005)
216. Marcialis, R.L., Rieke, G.H., Lebofsky, L.A.: The surface composition of Charon—tentative identification of water ice. Science **237**, 1349–1351 (1987)
217. Marsh, K.A., Dowell, C.D., Velusamy, T., Grogan, K., Beichman, C.A.: Images of the Vega dust ring at 350 and 450 μm: new clues to the trappingof multiple-sized dust particles in planetary resonances. Astrophys. J. Lett. **646**, L77–L80 (2006)
218. Masson, P., Carr, M.H., Costard, F., Greeley, R., Hauber, E., Jaumann, R.: Geomorphologic evidence for liquid water. Space Sci. Rev. **96**, 333–364 (2001)
219. Matese, J.J., Whitmire, D.P., Reynolds, R.T.: Dust clouds around red giant stars—evidence of sublimating comet disks? Icarus **81**, 24–30 (1989)
220. Matsui, T., Tajika, E., Abe, Y.: Climate and impact: climatic change on mars caused by impact basin formation. In: Lunar and Planetary Institute Conference Abstracts. Lunar and Planetary Inst., Technical Report, vol. 19, p. 742 (1988)
221. Mayor, M., Queloz, D.: A Jupiter-mass companion to a solar-type star. Nature **378**, 355 (1995)
222. Mayor, M., Udry, S., Lovis, C., Pepe, F., Queloz, D., Benz, W., Bertaux, J.-L., Bouchy, F., Mordasini, C., Segransan, D.: The HARPS search for southern extra-solar planets. XIII. A planetary system with 3 super-Earths (4.2, 6.9, and 9.2 M.). Astron. Astrophys. **493**, 639–644 (2009)
223. McConnochie, T.H., Buratti, B.J., Hillier, J.K., Tryka, K.A.: A search for water ice at the lunar poles with Clementine images. Icarus **156**, 335–351 (2002)
224. McDougall, J.D.: Science **239**, 485 (1988)
225. McKinnon, W.B., Barr, A.C.: On the stability of an ocean within Enceladus. In: Lunar and Planetary Institute Conference Abstracts, vol. 39, p. 2517 (2008)
226. Meijerink, R., Poelman, D.R., Spaans, M., Tielens, A.G.G.M., Glassgold, A.E.: Rotational line emission from water in protoplanetary disks. Astrophys. J. Lett. **689**, L57–L60 (2008)
227. Melosh, H.J., Ekholm, A.G., Showman, A.P., Lorenz, R.D.: Is Europa's subsurface water ocean warm? In: Lunar and Planetary Institute Conference Abstracts, vol. 33, p. 1824 (2002)
228. Melosh, H.J., Ekholm, A.G., Showman, A.P., Lorenz, R.D.: The temperature of Europa's subsurface water ocean. Icarus **168**, 498–502 (2004)
229. Menou, K., Cho, J.Y.-K., Seager, S., Hansen, B.M.S.: "Weather" variability of close-in extrasolar giant planets. Astrophys. J. Lett. **587**, L113–L116 (2003)
230. Miller-Ricci, E., Seager, S., Sasselov, D.: The atmospheric signatures of super-Earths how to distinguish between hydrogen-rich and hydrogen-poor atmospheres. Astrophys. J. **690**, 1056–1067 (2009)
231. Miner, E.D., Wessen, R.R.: Neptune. The Planet, Rings and Satellites. Springer, Berlin (2002)
232. Mitri, G., Showman, A.P., Lunine, J.I., Lopes, R.M.C.: Resurfacing of Titan by ammonia-water cryomagma. Icarus **196**, 216–224 (2008)
233. Mochizuki, N., Hachisuka, K., Umemoto, T.: Survey of Outer galaxy molecular lines associated with water masers. In: Hagiwara, Y., Fomalont, E., Tsuboi, M., Yasuhiro, M. (eds.) Astronomical Society of the Pacific Conference Series, vol. 402, p. 384 (2009)

234. Mojzsis, S.J., Arrhenius, G., McKeegan, K.D., Harrison, T.M., Nutman, A.P., Friend, C.R.L.: Evidence for life on Earth before 3,800 million years ago. Nature **384**, 55–59 (1996)
235. Moore, M.H., Hudson, R.L.: Infrared study of ion-irradiated N_2-dominated ices relevant to Triton and Pluto: formation of HCN and HNC. Icarus **161**, 486–500 (2003)
236. Moore, W.B.: Tidal heating and convection in Io. J. Geophys. Res. (Planets) **108**, 5096 (2003)
237. Moses, J.I., Rawlins, K., Zahnle, K., Dones, L.: External sources of water for Mercury's putative ice deposits. Icarus **137**, 197–221 (1999)
238. Moutou, C., Coustenis, A., Schneider, J., St Gilles, R., Mayor, M., Queloz, D., Kaufer, A.: Search for spectroscopical signatures of transiting HD 209458b's exosphere. Astron. Astrophys. **371**, 260–266 (2001)
239. Mueller-Wodarg, I., Mendillo, M., Moore, L.: Water on Saturn: global effects on ionospheric densities. In: AGU Fall Meeting Abstracts, p. 1304 (2006)
240. Mumma, M.: Session 12. Comets and primitive asteroids: their role in supplying water and prebiotic organics to early Earths. Astrobiology **8**, 339–343 (2008)
241. Mumma, M.J., Weaver, H.A., Larson, H.P., Williams, M., Davis, D.S.: Detection of water vapor in Halley's comet. Science **232**, 1523–1528 (1986)
242. Murchie, S.L., Mustard, J.F., Ehlmann, B.L., Milliken, R.E., Bishop, J.L., McKeown, N.K., Noe Dobrea, E.Z., Seelos, F.P., Buczkowski, D.L., Wiseman, S.M., Arvidson, R.E., Wray, J.J., Swayze, G., Clark, R.N., Des Marais, D.J., McEwen, A.S., Bibring, J.-P.: A synthesis of Martian aqueous mineralogy after 1 Mars year of observations from the Mars reconnaissance orbiter. J. Geophys. Res. (Planets) **114**, E00D06 (2009)
243. Mustard, J.F., Cooper, C.D., Rifkin, M.K.: Evidence for recent climate change on Mars from the identification of youthful near-surface ground ice. Nature **412**, 411–414 (2001)
244. Nakamura, T., Tajika, E.: Climate change of Mars-like planets due to obliquity variations: implications for Mars. Geophys. Res. Lett. **30**(13), 130000 (2003)
245. Nash, D.B.: On Io's 2.788-micron band: origin by SO_2 or H_2O? Icarus **107**, 418–421 (1994)
246. Neish, C.D., Somogyi, Á., Smith, M.A.: Titan's primordial soup: formation of amino acids via low-temperature hydrolysis of tholins. Astrobiology **10**, 337–347 (2010)
247. Nihei, T.C., Lehner, M.J., Bianco, F.B., King, S.-K., Giammarco, J.M., Alcock, C.: Detectability of occultations of stars by objects in the Kuiper belt and Oort cloud. Astron. J. **134**, 1596–1612 (2007)
248. Nisbet, E.G., Sleep, N.H.: The habitat and nature of early life. Nature **409**, 1083–1091 (2001)
249. Nixon, C.A., Jennings, D.E., de Kok, R., Coustenis, A., Flasar, F.M.: Water in Titan's atmosphere from Cassini CIRS observations. In: Bulletin of the American Astronomical Society, vol. 38, p. 529 (2006)
250. Noll, K., Stephens, D., Grundy, W., Cruikshank, D., Romanishin, W., Tegler, S.: Infrared photometry of Kuiper belt objects with NICMOS. In: Bulletin of the American Astronomical Society, vol. 37, p. 746 (2005)
251. Nyffenegger, P., Davis, D.M., Consolmagno, G.J.: Tectonic lineations and frictional faulting on a relatively simple body (Ariel). Planet. Space Sci. **45**, 1069–1080 (1997)
252. Oki, T., Kanae, S.: Global hydrological cycles and world water resources. Science **313**, 1068–1072 (2006)
253. Olive, K.A.: Primordial Big Bang nucleosynthesis. ArXiv Astrophysics e-prints (1999)
254. Oliver, B.M.: Rationale for the water hole. Acta Astronaut. **6**, 71–79 (1979)
255. Oort, J.H.: The structure of the cloud of comets surrounding the Solar System and a hypothesis concerning its origin. Astron. J. **11**, 91–110 (1950)
256. Owen, T.C., Roush, T.L., Cruikshank, D.P., Elliot, J.L., Young, L.A., de Bergh, C., Schmitt, B., Geballe, T.R., Brown, R.H., Bartholomew, M.J.: Surface ices and the atmospheric composition of Pluto. Science **261**, 745–748 (1993)
257. Oyama, K., Abe, T.: Detection of water group ions from Comet Halley by means of Sakigake. In: Comet Halley. Investigations, Results, Interpretations, vol. 1, pp. 169–180 (1990)
258. Paige, D.A., Wood, S.E., Vasavada, A.R.: The thermal stability of water ice at the poles of Mercury. Science **258**, 643–646 (1992)

259. Peimbert, M.: Planetary nebulae. Rep. Prog. Phys. **53**, 1559–1619 (1990)
260. Petit, J.-M., Morbidelli, A., Chambers, J.E., Lunine, J.I., Robert, F., Valsecchi, G.B., Cyr, K.E.: Asteroid belt clearing and delivery of water to Earth. In: Bulletin of the American Astronomical Society, vol. 32, p. 1100 (2000)
261. Pieters, C.M., Goswami, J.N., Clark, R.N., Annadurai, M., Boardman, J., Buratti, B., Combe, J., Dyar, M.D., Green, R., Head, J.W., Hibbitts, C., Hicks, M., Isaacson, P., Klima, R., Kramer, G., Kumar, S., Livo, E., Lundeen, S., Malaret, E., McCord, T., Mustard, J., Nettles, J., Petro, N., Runyon, C., Staid, M., Sunshine, J., Taylor, L.A., Tompkins, S., Varanasi, P.: Character and spatial distribution of OH/H_2O on the surface of the Moon seen by M^3 on Chandrayaan-1. Science **326**, 568 (2009)
262. Pilcher, C.B.: The stability of water on Io. Icarus **37**, 559–574 (1979)
263. Pinte, C., Ménard, F.: A model of the accretion disk around AA Tau. In: Beaulieu, J., Lecavelier Des Etangs, A., Terquem, C. (eds.) Extrasolar Planets: Today and Tomorrow. Astronomical Society of the Pacific Conference Series, vol. 321, p. 343 (2004)
264. Pitman, K.M., Buratti, B.J., Mosher, J.A., Bauer, J.M., Momary, T.W., Brown, R.H., Nicholson, P.D., Hedman, M.M.: First high solar phase angle observations of Rhea using Cassini VIMS: upper limits on water vapor and geologic activity. Astrophys. J. Lett. **680**, L65–L68 (2008)
265. Plescia, J.B.: Cratering history of the Uranian satellites—Umbriel, Titania, and Oberon. J. Geophys. Res. **92**, 14918–14932 (1987)
266. Pont, F., Gilliland, R.L., Knutson, H., Holman, M., Charbonneau, D.: Transit infrared spectroscopy of the hot Neptune around GJ 436 with the Hubble Space Telescope. Mon. Not. **393**, 6–10 (2009)
267. Prialnik, D., Merk, R.: Growth and evolution of small porous icy bodies with an adaptive-grid thermal evolution code. I. Application to Kuiper belt objects and Enceladus. Icarus **197**, 211–220 (2008)
268. Protopapa, S., Herbst, T., Böhnhardt, H.: Surface ice spectroscopy of Pluto, Charon and Triton. Messenger **129**, 58–60 (2007)
269. Queloz, D., Bouchy, F., Moutou, C., Hatzes, A., Hébrard, G., Alonso, R., Auvergne, M., Baglin, A., Barbieri, M., Barge, P., Benz, W., Bordé, P., Deeg, H.J., Deleuil, M., Dvorak, R., Erikson, A., Ferraz Mello, S., Fridlund, M., Gandolfi, D., Gillon, M., Guenther, E., Guillot, T., Jorda, L., Hartmann, M., Lammer, H., Léger, A., Llebaria, A., Lovis, C., Magain, P., Mayor, M., Mazeh, T., Ollivier, M., Pätzold, M., Pepe, F., Rauer, H., Rouan, D., Schneider, J., Segransan, D., Udry, S., Wuchterl, G.: The CoRoT-7 planetary system: two orbiting super-Earths. Astron. Astrophys. **506**, 303–319 (2009)
270. Rabinowitz, D.L., Barkume, K., Brown, M.E., Roe, H., Schwartz, M., Tourtellotte, S., Trujillo, C.: Photometric observations constraining the size, shape, and albedo of 2003 EL61, a rapidly rotating, Pluto-sized object in the Kuiper belt. Astrophys. J. **639**, 1238–1251 (2006)
271. Ramanathan, V., Crutzen, P.J., Kiehl, J.T., Rosenfeld, D.: Aerosols, climate, and the hydrological cycle. Science **294**, 2119–2124 (2001)
272. Rauer, H., Collier-Cameron, A., Barnes, J., Harris, A.W.: Search for signatures of an atmosphere of HD209458 b. In: Penny, A. (ed.) Planetary Systems in the Universe. IAU Symposium, vol. 202, p. 109 (2004)
273. Raulin, F.: Exo-astrobiological aspects of Europa and Titan: from observations to speculations. Space Sci. Rev. **116**, 471–487 (2005)
274. Raulin, F.: Astrobiology and habitability of Titan. Space Sci. Rev. **135**, 37–48 (2008)
275. Raup, D.M., Sepkoski, J.J.: Testing for periodicity of extinction. Science **241**, 94–96 (1988)
276. Rawlins, K., Moses, J.I., Zahnle, K.J.: Exogenic sources of water for Mercury's polar ice. In: Bulletin of the American Astronomical Society, vol. 27, p. 1117 (1995)
277. Reach, W.T., Rho, J.: Detection of far-infrared water vapor, hydroxyl, and carbon monoxide emissions from the supernova Remnant 3C 391. Astrophys. J. Lett. **507**, 93–97 (1998)
278. Reinhard, R.: Summary of the scientific results of the Giotto Mission to Halley's comet: nucleus, dust and gas composition and fluxes. In: Cometary and Solar Plasma Physics, p. 81 (1988)

279. Rollinde, E., Vangioni, E., Maurin, D., Olive, K.A., Daigne, F., Silk, J., Vincent, F.H.: Influence of population III stars on cosmic chemical evolution. Mon. Not. **398**, 1782–1792 (2009)
280. Roos-Serote, M., Atreya, S.K., Wong, M.K., Drossart, P.: On the water abundance in the atmosphere of Jupiter. Planet. Space Sci. **52**, 397–414 (2004)
281. Roques, F., Doressoundiram, A.: The Kuiper belt explorated by stellar occultations. In: Bulletin of the American Astronomical Society, vol. 37, p. 746 (2005)
282. Rostek, F., Ruhlandt, G., Bassinot, F.C., Muller, P.J., Labeyrie, L.D., Lancelot, Y., Bard, E.: Reconstructing sea surface temperature and salinity using $\delta^{18}O$ and alkenone records. Nature **364**, 319–321 (1993)
283. Roush, T.L., Noll, K., Cruikshank, D.P., Pendleton, Y.J.: UV spectra of the Uranian satellites, Ariel, Titania, and Oberon. In: Bulletin of the American Astronomical Society, vol. 29, p. 1010 (1997)
284. Roy-Poulsen, H., Larsen, L., Roy-Poulsen, N.O., Vistisen, L., Knudsen, J.M.: Meteorites and the origin of water and reduced carbon on the Earth. Phys. Scr. **23**, 1113–1117 (1981)
285. Rubincam, D.P.: Mars—change in axial tilt due to climate? Science **248**, 720 (1990)
286. Saal, A.E., Hauri, E.H., Cascio, M.L., van Orman, J.A., Rutherford, M.C., Cooper, R.F.: Volatile content of lunar volcanic glasses and the presence of water in the Moon's interior. Nature **454**, 192–195 (2008)
287. Sagan, C., Chyba, C.: The early faint sun paradox: organic shielding of ultraviolet-labile greenhouse gases. Science **276**, 1217–1221 (1997)
288. Sagan, C., Toon, O.B., Gierasch, P.J.: Climatic change on Mars. Science **181**, 1045–1049 (1973)
289. Saha, M.N.: Molecules in interstellar space? Nature **139**, 840 (1937)
290. Sajina, A., Yan, L., Spoon, H., Fadda, D.: Detection of the 3.3 μm PAH feature as well as water ice and HAC absorption in z~2 ULIRGs. ArXiv e-prints (2009)
291. Salama, F., Allamandola, L.J., Witteborn, F.C., Cruikshank, D.P., Sandford, S.A., Bregman, J.D.: The 2.5-5.0 micron spectra of Io—evidence for H_2S and H_2O frozen in SO_2. Icarus **83**, 66–82 (1990)
292. Salama, F., Allamandola, L.J., Sandford, S.A., Bregman, J.D., Witteborn, F.C., Cruikshank, D.P.: Is H_2O present on Io? The detection of a new strong band near 3590/cm (2.79 microns). Icarus **107**, 413–417 (1994)
293. Schaefer, B.E.: The heliacal rise of Sirius and ancient Egyptian chronology. J. Hist. Astron. **31**, 149–155 (2000)
294. Schenk, P.M., Zahnle, K.: On the negligible surface age of Triton. Icarus **192**, 135–149 (2007)
295. Schneider, N.M., Burger, M.H., Schaller, E.L., Brown, M.E., Johnson, R.E., Kargel, J.S., Dougherty, M.K., Achilleos, N.A.: No sodium in the vapour plumes of Enceladus. Nature **459**, 1102–1104 (2009)
296. Schorghofer, N., Taylor, G.J.: Subsurface migration of H_2O at lunar cold traps. J. Geophys. Res. (Planets) **112**, 2010 (2007)
297. Schubert, G., Ellsworth, K., Stevenson, D.J.: Ice in the interiors of ganymede and Callisto. In: Smith, J. (ed.) Planetary Water, p. 36 (1980)
298. Schubert, G., Hussmann, H., Lainey, V., Matson, D.L., McKinnon, W.B., Sohl, F., Sotin, C., Tobie, G., Turrini, D., van Hoolst, T.: Evolution of icy satellites. Space Sci. Rev., 54 (2010)
299. Schulz, R.: Trans-Neptunian objects. Astron. Astrophys. Rev. **11**, 1–31 (2002)
300. Schulz, R., Owens, A., Rodriguez-Pascual, P.M., Lumb, D., Erd, C., Stüwe, J.A.: Detection of water ice grains after the DEEP IMPACT onto Comet 9P/Tempel 1. Astron. Astrophys. **448**, 53–56 (2006)
301. Schulze-Makuch, D., Dohm, J.M., Fan, C., Fairén, A.G., Rodriguez, J.A.P., Baker, V.R., Fink, W.: Exploration of hydrothermal targets on Mars. Icarus **189**, 308–324 (2007)
302. Seager, S., Sasselov, D.D.: Theoretical transmission spectra during extrasolar giant planet transits. Astrophys. J. **537**, 916–921 (2000)
303. Seaquist, E.R., Ivison, R.J., Hall, P.J.: A study of OH and H_2O maser emission in Mira type symbiotic stars. Mon. Not. **276**, 867 (1995)

304. Sekine, Y., Imanaka, H., Matsui, T., Khare, B.N., Bakes, E.L.O., McKay, C.P., Sugita, S.: The role of organic haze in Titan's atmospheric chemistry. I. Laboratory investigation on heterogeneous reaction of atomic hydrogen with Titan tholin. Icarus **194**, 186–200 (2008)
305. Seth, A.C., Greenhill, L.J., Holder, B.P.: Water masers as tracers of protostellar disks and outflows in the intermediate-mass star-forming region NGC 2071. Astrophys. J. **581**, 325–334 (2002)
306. Shaviv, N.J.: Toward a solution to the early faint Sun paradox: a lower cosmic ray flux from a stronger solar wind. J. Geophys. Res. (Space Phys.) **108**, 1437 (2003)
307. Shenk, P.M., McKinnon, W.B., Gwynn, D., Moore, J.M.: Flooding of Ganymede's bright terrains by low-viscosity water-ice lavas. Nature **410**, 57–60 (2001)
308. Shiba, H., Sato, S., Yamashita, T., Kobayashi, Y., Takami, H.: Detection of water vapor in T Tauri stars. Astrophys. J. Suppl. **89**, 299–319 (1993)
309. Showman, A.P., Guillot, T.: Atmospheric circulation and tides of "51 Pegasus b-like" planets. Astron. Astrophys. **385**, 166–180 (2002)
310. Sicardy, B., Widemann, T., Colas, F., Lecacheux, J., Pau, S., Thuillot, W., Beisker, W., Birnbaum, C., Hubbard, W.B., Hill, R.E., Porto, J., López, E., Cósias, R., Pallo, E., Percz, R., Pulupa, D., Simbaña, X., Yajamín, A., Recalde, E.: The Occultation of Hip 106829 by Titania. In: Bulletin of the American Astronomical Society, vol. 33, p. 1130 (2001)
311. Simon-Miller, A.A., Conrath, B., Gierasch, P.J., Beebe, R.F.: A detection of water ice on Jupiter with Voyager IRIS. Icarus **145**, 454–461 (2000)
312. Sonnabend, G., Wirtz, D., Schieder, R., Bernath, P.F.: High-resolution infrared measurements of H_2O and SiO in sunspots. Sol. Phys. **233**, 205–213 (2006)
313. Sotin, C., Head, J.W., Tobie, G.: Europa: Tidal heating of upwelling thermal plumes and the origin of lenticulae and chaos melting. Geophys. Res. Lett. **29**(8), 080000 (2002)
314. Spencer, J.: Planetary science: Enceladus with a grain of salt. Nature **459**, 1067–1068 (2009)
315. Spencer, J.R.: Icy Galilean satellite reflectance spectra—less ice on Ganymede and Callisto? Icarus **70**, 99–110 (1987)
316. Spencer, J.R., Pearl, J.C., Segura, M., Flasar, F.M., Mamoutkine, A., Romani, P., Buratti, B.J., Hendrix, A.R., Spilker, L.J., Lopes, R.M.C.: Cassini encounters Enceladus: background and the discovery of a south polar hot spot. Science **311**, 1401–1405 (2006)
317. Spencer, J.R., Barr, A.C., Esposito, L.W., Helfenstein, P., Ingersoll, A.P., Jaumann, R., McKay, C.P., Nimmo, F., Waite, J.H.: Enceladus: an active cryovolcanic satellite. In: Dougherty, M.K., Esposito, L.W., Krimigis, S.M. (eds.) Saturn from Cassini-Huygens, pp. 683–724. Springer, Berlin, (2009)
318. Spohn, T., Schubert, G.: Oceans in the icy Galilean satellites of Jupiter? Icarus **161**, 456–467 (2003)
319. Sprague, A.L., Hunten, D.M., Hill, R.E., Rizk, B., Wells, W.K.: Martian water vapor, 1988–1995. J. Geophys. Res. **101**, 23229–23254 (1996)
320. Sprague, A.L., Hunten, D.M., Hill, R.E., Doose, L.R., Rizk, B.: Water vapor abundances over Mars north high latitude regions: 1996–1999. Icarus **154**, 183–189 (2001)
321. Squyres, S.W.: Topographic domes on Ganymede—ice vulcanism or isostatic upwarping. Icarus **44**, 472–480 (1980)
322. Squyres, S.W., Reynolds, R.T., Cassen, P.M.: Liquid water and active resurfacing on Europa. Nature **301**, 225 (1983)
323. Stacy, N.J.S., Campbell, D.B., Ford, P.G.: Arecibo radar mapping of the lunar poles: a search for ice deposits. Science **276**, 1527–1530 (1997)
324. Starukhina, L.V.: High radar response of Mercury polar regions: water ice or cold silicates? In: Lunar and Planetary Institute Science Conference Abstracts. Lunar and Planetary Inst., Technical Report, vol. 31, p. 1301 (2000)
325. Stern, S.A., Shull, J.M.: Water masers, red giants, and Oort clouds around other stars. LPI Contrib. **765**, 208 (1991)
326. Stevenson, D.J., Fishbein, E.: The behavior of water in the giant planets. In: Lunar and Planetary Institute Conference Abstracts, vol. 12, pp. 1040–1042 (1981)
327. Storrs, A., Eney, B.: Synoptic observations of 134340 Pluto from the Hubble Space Telescope. In: Bulletin of the American Astronomical Society, vol. 41, p. 456 (2010)

328. Stull, M.A.: International legal protection of the "Water Hole" frequency band for a search of extraterrestrial intelligent life. In: Bulletin of the American Astronomical Society, vol. 7, p. 440 (1975)
329. Suárez, O., Gómez, J.F., Morata, O.: New detections of H_2O masers in planetary nebulae and post-AGB stars using the Robledo-70 m antenna. Astron. Astrophys. **467**, 1085–1091 (2007)
330. Svetsov, V.V.: Impact erosion of water ocean on the early Earth with a thin atmosphere. In: Lunar and Planetary Institute Science Conference Abstracts, vol. 40, p. 1147 (2009)
331. Swings, P., Rosenfeld, L.: Considerations regarding interstellar molecules. Astrophys. J. **86**, 483–486 (1937)
332. Taylor, F.W., Coustenis, A.: Titan in the Solar system. Planet. Space Sci. **46**, 1085–1097 (1998)
333. Le Gall, A., Janssen, M.A., Paillou, P., Lorenz, R.D., Wall, S.D. (the Cassini Radar Team): Radar-bright channels on Titan. Icarus **207**, 948–958 (2010)
334. Thomas, N., Alexander, C., Keller, H.U.: Loss of the surface layers of comet nuclei. Space Sci. Rev., **138**, 165–177 (2008)
335. Thompson, A.B.: Water in the Earth's upper mantle. Nature **358**, 295–302 (1992)
336. Tielens, A.G.G.M.: Interstellar depletions and the life cycle of interstellar dust. Astrophys. J. **499**, 267 (1998)
337. Tobie, G., Čadek, O., Sotin, C.: Solid tidal friction above a liquid water reservoir as the origin of the south pole hotspot on Enceladus. Icarus **196**, 642–652 (2008)
338. Tokar, R.L., Wilson, R.J., Johnson, R.E., Henderson, M.G., Thomsen, M.F., Cowee, M.M., Sittler, E.C., Young, D.T., Crary, F.J., McAndrews, H.J., Smith, H.T.: Cassini detection of water-group pick-up ions in the Enceladus torus. Geophys. Res. Lett. **35**, 14202 (2008)
339. Treiman, A.H., Lanzirotti, A., Xirouchakis, D.: Ancient water on asteroid 4 Vesta: evidence from a quartz veinlet in the Serra de Magé eucrite meteorite. Earth Planet. Sci. Lett. **219**, 189–199 (2004)
340. Trilling, D.E., Bernstein, G.M.: Light curves of 20–100 km Kuiper Belt objects using the Hubble Space Telescope. Astron. J. **131**, 1149–1162 (2006)
341. Trujillo, C.A., Jewitt, D.C., Luu, J.X.: The scattered Kuiper Belt objects. In: AAS/Division for Planetary Sciences Meeting Abstracts, vol. 31, p. 26.02 (1999)
342. Tsuji, T.: Water in emission in the infrared space observatory spectrum of the Early M supergiant star μ Cephei. Astrophys. J. Lett. **540**, L99–L102 (2000)
343. Turtle, E.P., Perry, J., McEwen, A.S., Barbara, J., Del Genio, A.D., West, R.A., Hayes, A.G.: Cassini ISS observations of seasonal changes in Titan's meteorology and surface features (Invited). In: AGU Fall Meeting Abstracts (2009)
344. Valeille, A., Bougher, S.W., Tenishev, V., Combi, M.R., Nagy, A.F.: Water loss and evolution of the upper atmosphere and exosphere over martian history. Icarus **206**, 28–39 (2010)
345. van Hoolst, T., Rambaux, N., Karatekin, Ö., Dehant, V., Rivoldini, A.: The librations, shape, and icy shell of Europa. Icarus **195**, 386–399 (2008)
346. Visscher, C., Fegley, B.J.: Chemical constraints on the water and total oxygen abundances in the deep atmosphere of Saturn. Astrophys. J. **623**, 1221–1227 (2005)
347. Vogt, S.S., Wittenmyer, R.A., Butler, R.P., O'Toole, S., Henry, G.W., Rivera, E.J., Meschiari, S., Laughlin, G., Tinney, C.G., Jones, H.R.A., Bailey, J., Carter, B.D., Batygin, K.: A super-Earth and two Neptunes orbiting the nearby sun-like star 61 Virginis. Astrophys. J. **708**, 1366–1375 (2010)
348. Vourlidas, A., Davis, C.J., Eyles, C.J., Crothers, S.R., Harrison, R.A., Howard, R.A., Moses, J.D., Socker, D.G.: First direct observation of the interaction between a comet and a coronal mass ejection leading to a complete plasma tail disconnection. Astrophys. J. Lett. **668**, L79–L82 (2007)
349. Wagner, R.J., Neukum, G., Giese, B., Roatsch, T., Denk, T., Wolf, U., Porco, C.C.: Geology of Saturn's satellite Rhea on the basis of the high-resolution images from the targeted flyby 049 on Aug. 30, 2007. In: Lunar and Planetary Institute Conference Abstracts, vol. 39, p. 1930 (2008)

350. Waite, J.H. Jr., Lewis, W.S., Magee, B.A., Lunine, J.I., McKinnon, W.B., Glein, C.R., Mousis, O., Young, D.T., Brockwell, T., Westlake, J., Nguyen, M., Teolis, B.D., Niemann, H.B., McNutt, R.L., Perry, M., Ip, W.-H.: Liquid water on Enceladus from observations of ammonia and ^{40}Ar in the plume. Nature **460**, 487–490 (2009)
351. Wall, S., Hayes, A., Bristow, C., Lorenz, R., Stofan, E., Lunine, J., Le Gall, A., Janssen, M., Lopes, R., Wye, L., Soderblom, L., Paillou, P., Aharonson, O., Zebker, H., Farr, T., Mitri, G., Kirk, R., Mitchell, K., Notarnicola, C., Casarano, D., Ventura, B.: Active shoreline of Ontario Lacus, Titan: a morphological study of the lake and its surroundings. Geophys. Res. Lett. **37**, 5202 (2010)
352. Wallace, L., Bernath, P., Livingston, W., Hinkle, K., Busler, J., Guo, B., Zhang, K.: Water on the Sun. Science **268**, 1155 (1995)
353. Walsh, A.J., Lo, N., Burton, M.G., White, G.L., Purcell, C.R., Longmore, S.N., Phillips, C.J., Brooks, K.J.: A pilot survey for the H_2O southern galactic plane survey. Publ. Astron. Soc. Aust. **25**, 105–113 (2008)
354. Watson, K., Murray, B., Brown, H.: On the possible presence of ice on the Moon. J. Geophys. Res. **66**, 1598 (1961)
355. Weaver, H., Williams, D.R.W., Dieter, N.H., Lum, W.T.: Observations of a strong unidentified microwave line and of emission from the OH molecule. Nature **208**, 29 (1965)
356. Weinreb, S., Barrett, A.H., Meeks, M.L., Henry, J.C.: Radio observations of OH in the interstellar medium. Nature **200**, 829 (1963)
357. Wild, M., Liepert, B.: EDITORIAL: The Earth radiation balance as driver of the global hydrological cycle. Environ. Res. Lett. **5**(2), 025203 (2010)
358. Williams, D.M., Zugger, M.: Polarization of starlight by exoplanets with watery surfaces. In: Bulletin of the American Astronomical Society, vol. 41, p. 328 (2010)
359. Williams, D.M., Kasting, J.F., Wade, R.A.: Habitable moons around extrasolar giant planets. Nature **385**, 234–236 (1997)
360. Wilson, C.D., Booth, R.S., Olofsson, A.O.H., Olberg, M., Persson, C.M., Sandqvist, A., Hjalmarson, Â., Buat, V., Encrenaz, P.J., Fich, M., Frisk, U., Gerin, M., Rydback, G., Wiklind, T.: Upper limits to the water abundance in starburst galaxies. Astron. Astrophys. **469**, 121–124 (2007)
361. Wilson, L., Head, J.W.: Europa cryovolcanism: ascent and eruption of magma and its role in resurfacing. In: Lunar and Planetary Institute Conference Abstracts, vol. 29, p. 1138 (1998)
362. Winnberg, A., Engels, D., Brand, J., Baldacci, L., Walmsley, C.M.: Water vapour masers in long-period variable stars. I. RX Bootis and SV Pegasi. Astron. Astrophys. **482**, 831–848 (2008)
363. Wood, S.E., Vasavada, A.R., Paige, D.A.: Temperatures in the polar regions of Mercury: implications for water ice. In: Bulletin of the American Astronomical Society, vol. 24, p. 957 (1992)
364. Woodall, J.M., Gray, M.D.: A search for 22-GHz H_2O masers in supernova remnants. Mon. Not. **378**, L20–L23 (2007)
365. Woolf, N.J., Schwarzschild, M., Rose, W.K.: Infrared spectra of red-giant stars. Astrophys. J. **140**, 833 (1964)
366. Woosley, S.E., Heger, A., Weaver, T.A.: The evolution and explosion of massive stars. Rev. Mod. Phys. **74**, 1015–1071 (2002)
367. Wordsworth, R., Forget, F., Millour, E., Madeleine, J.-B., Eymet, V., Haberle, R.: Three-dimensional modelling of the early Martian climate and water cycle. In: Lunar and Planetary Institute Science Conference Abstracts. Lunar and Planetary Inst., Technical Report, vol. 41, p. 1913 (2010)
368. Yang, B., Jewitt, D.: Spectroscopic search for water ice on Jovian Trojan asteroids. Astron. J. **134**, 223–228 (2007)
369. Yang, F., Kumar, A., Schlesinger, M.E., Wang, W.: Intensity of hydrological cycles in warmer climates. J. Climate **16**, 2419–2423 (2003)
370. Yokohata, T., Odaka, M., Kuramoto, K.: Role of H_2O and CO_2 ices in martian climate changes. Icarus **159**, 439–448 (2002)

371. Zajac, B., Ingersoll, A.P., Dowling, T.E.: Giant planet formation using shallow water simulations of a gaseous protostellar disk. In: Bulletin of the American Astronomical Society, vol. 23, p. 1231 (1991)
372. Zhong, F., Mitchell, K.L., Hays, C.C., Choukroun, M., Barmatz, M., Kargel, J.S.: The rheology of cryovolcanic slurries: motivation and phenomenology of methanol-water slurries with implications for Titan. Icarus **202**, 607–619 (2009)
373. Zhou, J.-L.: Formation and tidal evolution of hot super-Earths in multiple planetary systems. ArXiv e-prints (2009)
374. Zimmerer, K.S.: The origins of Andean irrigation. Nature **378**, 481–483 (1995)
375. Zobov, N.F., Polyansky, O.L., Tennyson, J., Shirin, S.V., Nassar, R., Hirao, T., Imajo, T., Bernath, P.F., Wallace, L.: Using laboratory spectroscopy to identify lines in the K- and L-band spectrum of water in a sunspot. Astrophys. J. **530**, 994–998 (2000)
376. Zolotov, M.Y., Fegley, B., Lodders, K.: Hydrous silicates and water on Venus. Icarus **130**, 475–494 (1997)

Index

A. Hanslmeier, *Water in the Universe*, Astrophysics and Space Science Library 368,
DOI 10.1007/978-90-481-9984-6, © Springer Science+Business Media B.V. 2011

MIX
Papier aus verantwortungsvollen Quellen
Paper from responsible sources
FSC® C105338

If you have any concerns about our products,
you can contact us on
ProductSafety@springernature.com

In case Publisher is established outside the EU,
the EU authorized representative is:
Springer Nature Customer Service Center GmbH
Europaplatz 3, 69115 Heidelberg, Germany

Printed by Libri Plureos GmbH
in Hamburg, Germany